Technology, Innovation and Economic Policy

Edited by
Peter Hall
University of New England

Phili Allan

First published 1986 by
PHILIP ALLAN PUBLISHERS LIMITED
MARKET PLACE
DEDDINGTON
OXFORD OX5 4SE

British Library Cataloguing in Publication Data

Technology, innovation and economic policy
   1. Technological innovation − Economic aspects
   I. Hall, Peter, 1948−
   338'.06   HC79.T4
ISBN 0-86003-062-8
ISBN 0-86003-171-3 Pbk

Printed and bound in Great Britain by The Camelot Press, Southampton

These papers are reprinted from the *Greek Economic Review*

CONTENTS

LIST OF CONTRIBUTORS

Jock R. Anderson is Professor of Agricultural Economics at the University of New England in Armidale, New South Wales.

A.S. Bhalla is Chief of the Technology and Employment Branch of the Employment and Development Department at the International Labour Office, Geneva.

John Enos is Fellow in Economics at Magdalen College, Oxford.

Peter Hall is Senior Lecturer in Economics at the University of New England in Armidale, New South Wales.

Peter Hazell is Research Fellow and Director, Development Strategy Program, International Food Policy Research Institute, Washington D.C.

Jeffrey James is Professor of Economics at Boston University, Massachusetts.

Don Lamberton is Professor of Economics at the University of Queensland.

Michèle Ledić is Senior Lecturer in Economics at the University of Zagreb.

Stuart Macdonald is Senior Lecturer in Economics at the University of Queensland.

Tom Mandeville is Lecturer in Economics at the University of Queensland.

J.S. Metcalfe is Professor of Economics at the University of Manchester.

Roy Rothwell is Senior Fellow at the Science Policy Research Unit, University of Sussex, Brighton.

Aubrey Silberston is Professor of Economics at Imperial College, London.

Sir Bruce Williams is Director of the Technical Change Centre, London.

INTRODUCTION

These essays have been brought together to give some idea of the diversity of current research into the process of technological innovation and the policies appropriate for its stimulation and direction. The collection first appeared in a special number of the *Greek Economic Review*.

The first essay attempts to provide an overview of developments in the area to date and may be viewed as a backdrop to all that follows. Metcalfe's contribution integrates the three mechanisms of diffusion, selection and inducement in an analysis which highlights the interaction between a technology and the economic environment in which it evolves. This sort of analysis is an important advance on earlier work in which a technology was viewed as diffusing in an otherwise unchanging economic setting. Rothwell argues that the worldwide economic problems of the last decade have their roots in the paucity of radical product innovations which could have regenerated demand growth at a time when older industries were simultaneously reaching maturity. If this analysis is correct, he suggests, this would indicate technology-oriented reindustrialisation policies to rekindle global growth. One of the more obvious features of the economic slowdown of recent years has been unemployment and in his paper, Williams surveys the views of the many economists, classical and modern, who have tried to trace the connections between technological change and the employment of labour. Along the way, he discusses the 'long cycles' which are implicit in Rothwell's work.

Given the extent to which technological dominance is said to benefit the international competitiveness of nations, it is helpful to have in the essay by Ledić and Silberston an analysis of the way in which technology flows have been reflected in the balance of payments figures of different countries. One of their findings, unsurprisingly, is that developing countries pay large amounts to the developed countries for the use of technology inputs, partly because of the presence of subsidiaries within their borders of US, European and Japanese firms. This raises the question of how much in principle developing countries might gain from using new technology — an issue taken up by Bhalla and James in the first of a block of three essays in the volume with developmental significance. Bhalla and James point out some of the problems of integrating traditional and newly emerging technologies, and wonder how much hope there is of applying LDCs' recent advances in biotechnology, microelectronics and solar energy use. Enos's essay is a nice counterpoint to this. He argues that countries must choose between absorbing their modes of production from abroad or creating indigenous modes. For governments, an interesting question is to know how to allocate the scarce technical and administrative resources efficiently between these two competing activities. Enos suggests an answer to the problem in the framework of a game-theoretical model. In the last of these 'developmentally flavoured' essays Hazell and Anderson express concern that while the application of technology has permitted impressive agricultural growth rates since the Second World War, poverty is still widespread because of continuing population growth. This is a theme familiar in economics almost since its emergence as a science, but Hazell and Anderson believe that disseminating new technology is still the key to success and suggest policy guidelines to assist in this task.

The final essay in the volume attempts to open a window on the operation of a university research unit working on information technology. There is an element of 'the play within a play here': Lamberton, Macdonald and Mandeville strive to shed light upon their own strategy and problems, and at the same time illuminate the material upon which they have been working.

While the diversity of research methods in this area is well represented in this volume, a recurring theme of much recent work — evolutionary process — can also be found throughout. Evolutionary insights are central to Metcalfe's approach but also have relevance to Rothwell's views on the consequences of technological maturation and Williams's discussion of the effects of 'long cycles'. Again, existing technology-related flows in the balance of payments reflect the heritage of patterns of international technological strength. And for developing countries and advanced nations alike, future directions are inescapably constrained (in important measure) by the achievements and failures of past and present.

These essays make no pretence of 'covering the field' but they do give a flavour of the sort of questions being asked by researchers who have the nature and process of technological innovation as a central interest. At a time when the wellsprings of growth and productivity are again under scrutiny it is hoped that these essays will make a constructive contribution both to understanding and to policy making.

P. H. Hall

CHAPTER ONE

THE THEORY AND PRACTICE OF INNOVATION POLICY: AN OVERVIEW

By P.H. Hall*

This collection of essays is about innovation: what it is, what its effects are, and with an emphasis on how policy-makers should respond to it. Broadly speaking, innovations take two forms: product innovations and process innovations. The former involve changes in the specification of goods and services sold in the market, either as products for final, consumer demand or as products used as intermediate inputs in other parts of the economy. Process innovations involve changes in the nature of inputs or the way in which they are used in any given production process. As is clear, the distinction between product and process innovation is blurred: the product innovation of industry A (a producer, say, of materials) is simultaneously a process innovation for industries B, C,... which use that material as an input.

The innovation process refers to a very broad spectrum of activities. Scientists involved in basic research extend and deepen the pool of knowledge relating to natural and physical laws. When such research is undertaken with the expectation that it will help solve a known practical problem, basic research is strategic; when it aims merely to expand general understanding, without any deliberate or explicit commitment to practical problem solving, basic research is defined as pure. Applied research has a more specific focus on acquiring new knowledge to solve particular problems and experimental development draws on basic and applied research to produce new materials, products and devices or install new processes. The innovation process then, involves both creating new knowledge and drawing on the knowledge pool to generate new products and processes. Further, the innovation process extends beyond the initial introduction of a new product to its diffusion among potential consumers and/or users and includes the responses of producers to market feed-back from buyers of new and existing products.

Often, the innovation process is divided into three parts: invention - devising a new product or process: innovation - actually introducing the invention to the market for the first time: diffusion - subsequent production and consumption of the invention through the economy. Again, the technology may be taken to describe all known ways of converting inputs into out-

*University of New England, New South Wales, Armidale, Australia.

puts; technological change as finding new products or new ways of combin
ing inputs to make new or existing goods; techniques as "blueprints" for
converting inputs into outputs; and technical change as the introduction
of a process or product for the first time, and its subsequent diffusion.

Interest in innovation and technological change lies in the potential
they offer for welfare gain and the structural and adjustment problems they
inevitably bring with them. The growth in the potential for social welfare,
based on consumption, depends over time on the growth of productive
potential, real incomes and the distribution of income. Economic growth
springs, logically, from growth in the quantity of productive inputs, improvements in their quality and increases in the efficiency with which inputs are
combined. Design and material innovations, as examples, impinge directly
on the quality of inputs, being embodied in the inputs themselves. Process
innovation can also have implications for the more efficient combination of
inputs. (The substitution of electric for steam power, for example, permitted
the more efficient disposition of manufacturing equipment on shop floors).
Whether growth derives dominantly from quantity changes in inputs or
technological change has been a matter of great controversy and depends to
a large extent upon definitions and measurement methods used. After an
exhaustive analysis of this debate, however, Usher, (1980, p. 289), a leading
authority on growth accounting concluded: "... economic growth depends
predominantly on technical change and cannot occur to any significant extent
in its absence".

For policy makers interested in growth, this is good news - at least
to the extent that such conclusions seem to provide a focus for policy activity. But it is only a starting point. No growth accounting can do justice to
the changes in the composition of output which innovation brings to growth
and so questions arise as to the most desirable menu or mix of products.
If governments are interested at all in growth, they have to be interested in
the nature of the path it takes as well as its pace. Again different combinations and permutations of innovation may have widely varying implications
for other policy goals, such as employment, inflation, the balance of payments, and income distribution. In relation to the last of these, there can be
no guarantee that any arbitrary growth path will allow all to benefit equally
from the fruits of growth. Conflicts such as these will be taken up again later.

Acknowledging, however, that growth has the potential to benefit all,
what efforts if any should government make to promote it through encouraging the workings of innovation? The hard-nosed answer is "leave it to the
market" — but as the first part of this essay reminds us, there are good
reasons in this area to expect the market to generate welfare sub-optimal

solutions. To the question "Why intervene?" we can thus start by answering that the market is unreliable. An equally important matter is "Can governments do better than the (imperfect) market?" — and that turns out to be more difficult to resolve.

Assuming any case at all can be made for intervention, the next question is how to go about the task? Appropriate intervention calls for an understanding of how the innovation process works and what its effects will be. What is noted in this essay is that the innovation process is much more a network of sequential and simultaneous interactions than a unidirectional flow. Appropriate intervention is, therefore, far from simple and the record suggests that much of it may, in the past, have been misplaced or ineffective. In principle intervention may occur with the aim of stimulating or directing basic research or applied, experimental development, innovation or diffusion — i.e. any part of the innovation process. This explains why there are arguments for subsidising research in universities and government laboratories, for example, or investment subsidies directed towards the purchase of innovatory plant. Policy-makers should also be concerned to strengthen links and speed information flows between parts of the process known to be connected. This suggests for example policies directed towards speeding the diffusion of new techniques by easing the processes of technology transfer. How effective each policy instrument turns out to be will be one determinant of how many resources are directed to each one.

Why intervene?

To focus thoughts, recall first the two central propositions of welfare economics. (1) : Assuming households and firms act perfectly competitively (i.e. are all price-takers), that there is a complete set of markets for all goods and services and there is perfect information, then any competitive equilibrium is also a Pareto Optimum (P-O). (A P-O is defined as a situation in which no individual's utility level can be increased without decreasing that of another individual.) (2) : Assuming convex household indifference maps and firms' production sets, comprehensive markets, perfect information and the capacity costlessly to make lump sum taxes and subsidies, then for every P-O there are associated systems of prices and resource ownership which would achieve that P-O as a competitive equilibrium. These propositions can be shown to hold for "one-period economies" but can also be generalised to apply to any number of periods if there is no uncertainty about future tastes and technology and a complete set of markets permitting bargains to be struck in relation to transactions which will not occur until some future date. The equivalence of the competitive equilibrium and the P-O is reflected in the fact that in each case the marginal rate of substitution between any pair of goods is the same for each household, equal to the ratio of prices for those goods and in addition to the ratio of their marginal costs in production, the marginal rate of transformation. For a P-O, each household's MRS between every pair of goods must be equal to society's marginal rate of transformation between each corresponding pair, otherwise a welfare-superior allocation of resources could be found. A competitive equilibrium under the assumptions stated guarantees the achievement of these equalities. A P-O will not be achieved in the market if the stated assumptions are violated (market failure) but may be achieved by means other than competition (e.g. government dictat).

Market failure may arise for a variety of reasons. Emphasis is given here to cases related to the innovation process and its associated activities and effects[1].

Notice first that a necessary (though not sufficient) condition for technological change is inventive activity. Such activity may be undertaken for its own sake or under the inducement of perceived market need. Suppose that an inventor (an individual or a firm) has an innovative design or process and attempts to sell this information or knowledge in the marketplace. It almost immediately becomes clear that a market in knowledge of this kind is going to be difficult to construct. The reason for this is that potential buyers in the market cannot bid for new knowledge at a price reflecting their valuation of it unless they have been told what the knowledge comprises. Once they have been told, however, they then possess the knowledge and it would be irrational for them to offer to pay for it. In the case of physical goods, the seller need not relinquish any right to ownership or use of the commodity while potential buyers weigh up its value to them and until payment (or promise of payment) has been made. In the case of knowledge, the seller can only demonstrate the value of his wares by sharing his ideas, but since this automatically allows possession to pass (jointly) into the hands of the potential buyer, the seller reduces the marketable value of his wares in the process, perhaps to zero. This line of argument suggests there may be a threat even to the existence of the market structure whose operation is required to achieve an optimum. More specifically, it is also clear that if inventors cannot appropriate much or any of the value that society would gain from their ideas, they will react in ways which threaten a potential optimum. Firms may allocate less to R & D than they otherwise might. Individual inventors and small firms may try to develop commercially new ideas which

1. Arrow (1962).

it would have been less wasteful of resources to leave to other or larger concerns more experienced and skilled in the relevant engineering and management areas.

A central area of policy concern, therefore, has always been to devise some way of rewarding inventive activity lest the inappropriability problem outlined above discourage it or allow it to be misdirected. This has been seen as especially relevant in relation to pure research where applications are particularly diffuse and the opportunities for appropriating rewards neglible.

In addition to the appropriability problem Arrow notes that uncertainty and the indivisibility of information may threaten optimal solutions. First, uncertainty. The distribution of anticipated returns on an investment in researching and developing a new product or process will very often have a much greater variance for any given mean than would investing in more equipment of the kind that a firm already uses and knows. This greater variance may arise from the inherent unpredictability of how well and how fast research and/or marketing will proceed. It is also often related to the fact that a firm's competitors may be doing similar research. The outcome of the race among competitors is, however, uncertain: neither the date signifying the end of the race, nor the winner is known at the time when resources have initially to be committed. The greater variance in the distribution of expected returns will make firms, in the main assumed risk averse, shy away from innovatory projects when compared with those associated with greater uncertainty.

In an argument with relevance both to uncertainty and inappropriability, there may only be too little R & D investment *relative* to other kinds of investment[2]. If the overall amount of resources devoted to investment compared with consumption has been stimulated by macro-economic policy, the uncertainty argument may not be enough, by itself, to establish that R & D expenditure is sub-optimal. On the other hand, uncertainty often does penalise firms seeking external finance for their innovatory projects. It may be almost impossible for new or small firms to induce potential share-holders to buy an interest in their business — a capital market imperfection which has been widely noted[3]. But even for larger and well-established firms there may be reason to expect investment in R & D to be sub-optimal. This may simply be because the size of the investment or the length of project gestation discourages all but the most risk-loving investors (often a minority) to buy or retain an interest in the company.

2. Phelps (1965), Ch. 4.
3. Hay & Morris (1978), Ch. 17.

It can be argued that the greater uncertainty of returns associated with innovating firms ought not to discourage rational shareholders from holding their stock. Portfolio diversification ought to enable shareholders to achieve any degree of risk they desire for themselves. In fact, the transaction costs associated with floating new issues and the reluctance of managements to expose themselves to capital market discipline may often mean that R & D will be financed out of retained profits. When this happens, managers must take responsibility for the consequences, the stock market not having been informed and not knowingly having accepted any of the risk. If managers are risk averse, they will again eschew riskier projects. Even if it were possible for managers to shift risk, hovever, increased efficiency in risk-bearing would almost certainly lead to decreased technical efficiency[4]. The reason here is "moral hazard". The less managers are made responsible for the consequences of their actions, the less they are likely to strive for success.

While resolving capital market imperfections is thus indicated as another policy concern, it has to be recognised that the more the market at large helps bear risks, the less sure may be the success of the project. The best that can be hoped for is that moral hazard can be economised in shifting and reducing risk[5]. Given that uncertainty arises in part from firms' lack of information about their competitors' innovation strategies, there may also be a role for government in co-ordinating research efforts.

Other issues surround the indivisibility of a piece of knowledge or information about an innovation. When a product is indivisible — such as a bridge or a railway carriage — it is possible to allow increases in the number of persons consuming the services rendered by the good without any additional costs being incurred (at least up to the relevant capacity constraint). This phenomenon of nonrivalness in consumption is a feature of public goods, since by definition public goods are such that the addition of further consumers will not lead subtraction from any other individual's consumption of the good[6]. The point is relevant here because knowledge, once produced, can be shared by as many as wish to acquire it at no variable cost greater than the negligible cost of transmission. If price is to be set equal to marginal cost, all knowledge should on these grounds be (almost) freely available. To the extent that devices *can* be found (contrary to the appropriability arguments) to charge for new technological information, their success implies a divergence from the wel-

---

4. Arrow (op. cit.).
5. Demsetz, in Lamberton (1971), p. 168.
6. Samuelson (1954).

fare optimum. Demsetz[7] has argued here that it is confusing to separate the acts of producing and disseminating information in the way that underlies this point. Since a major *function* of paying a positive price is to encourage investment in R & D, the efficiency with which information is used cannot be judged *without* examining the effects on willingness to do research. A separate point, also related to indivisibility, is that it introduces the possibility that production possibility sets will be non-convex. This in turn means that the existence of a competitive equilibrium is not assured − in which case it becomes more difficult either to define or attain a welfare optimum[8].

As a closing note on the quantity of research, it is perhaps worth noting that a few theoretical results[9] exist which indicate what, in highly stylized growth models, the optimal rate of technological change might be. A particularly interesting finding is that if technological progress is endogenous (i.e. determined within the model, rather than falling like manna from heaven), firms must know all future prices in order to perform in a way which guarantees an efficient use of resources over the future growth of the economy. Given the absence of comprehensive future markets, this is a quite infeasible requirement. On the other hand. it can also be shown that if firms know and treat as parametric the Kennedy-Weiszacker relationship between rates of labour and capital-augmenting innovation, then if all firms are alike, the government can operate to produce an optimal amount of technological change and all the other conditions for an optimum will be achieved by private sector profit-maximizing[10]. It must be said that these results are derived from models so stylized in their approach that their relevance to policymakers is very much in question. In particular the models are built on the presupposition of equilibrium (steady state) growth paths and thus abstract from most of the interesting problems, of a disequilibrium nature, which actually arise when technological progress occurs[11].

Not all of the analytical work on technology policy has been performed in an equilibrium framework, however. One dimension of the problem, not explicitly considered thus far, is the optimal *mix* of innovation projects, and Nelson and Winter (NW)[12] have attacked this issue with an explicitly disequi-

7. Demsetz (op. cit.), pp. 171-2.

8. The non-convexity arises from the fact that a given piece of technological knowledge, once acquired, may be applied to any number of units of output without requiring supplementation. The general problem is explored by Arrow and Hahn (1971) pp. 60-2.

9. Nordhaus (1969), Ch. 6, Uzawa (1965), Phelps (1966).

10. Nordhaus (op. cit.).

11. Bliss (1975), p. 11.

12. Nelson & Winter (1982).

librium perspective motivated by pioneers such as Schumpeter and Simon[13].

In the NW vision[14], the rationality of firms' decision-makers is "bounded" in the sense that they do not know and cannot take account of all the complex repercussions of their actions in a highly interdependent and uncertain world. Within this general framework, and given the existence of an inherited technology on one hand and a yet-to-be discovered set of techniques on the other, it makes sense to characterize technological progress as the outcome of search activity performed in the "neighbourhood" of existing knowledge. This builds on the idea that "knowledge of a given technology is particularly transferable to related ones"[15]. Search is motivated partly by the prospect of profit (although NW-type firms are specifically not profit-maximizers but, rather, profit-seekers) and partly by competitive threat.

A result of this analysis of particular potential interest to policy makers is that, left to its own devices, the market is not likely to generate an optimal pattern (mix), or sequence, of technological change. It is reasonable to assume that each industry comprises firms which know roughly the direction in which to go to search for best practice techniques among the general kind already in use. If techniques close to best practice are unprotected by patent, firms are likely to converge upon and cluster around the same opportunity. NW are concerned that neighbourhood search and the potential for imitation may lead to too much concentration on some innovation probabilities and not enough on others. "If a firm explores new terrain, it is less likely to come up with something. And if it does, it knows that other firms will soon cluster around"[16].

Another implication of the NW analysis is that a scramble of the kind they describe may also generate innovation which is socially premature : the *timing* dimension is thus an additional issue. In a rigorous (and equilibrium) framework, Barzel[17] has shown generally that the socially optimal innovation date occurs when the marginal social benefits from earlier innovation equal the marginal R & D costs of drawing the innovation date forward. An interesting feature of the analysis is that it suggests a set of circumstances in which *too many* (rather than insufficient) resources are applied to R & D. These conditions may be stylized as a race in which many firms compete to win a wholly protective patent as the "reward" for being first to solve a

13. Schumpeter (1934) and Simon (1955 and 1959)
14. Nelson & Winter (op. cit.), p.35.
15. Nelson & Winter (op. cit.), p. 249.
16. Nelson & Winter (op. cit.), p. 388.
17. Barzel (1968).

given technological problem While the stylization has been criticized[18], it is easy to see, intuitively, how the incentive to be "first past the post" may lead to overinvestment in R & D. The innovation date determined in the market has been the subject of intense speculation[19] and turns out to depend upon the quality of R & D expenditure in relation to a particular innovation, market structure and degree of rivalry in the innovating industry, and size of innovation. (It also depends upon the degree to protection bestowed by devices such as patents, but this question will be taken up when policy remedies are considered). There is no guarantee that these factors can always (if ever) be relied upon to generate a socially optimal introduction date.

It perhaps ought to be added that if the scope exists for individual innovations taken in isolation to be introduced too early or too late, then there must also be the danger of non-optimal *sequences* of innovation occurring, a question which becomes important when a choice must be made among innovatory infrastructures upon which subsequent developments will be built.

While innovation may itself be the focus for policy, it should not be forgotten that innovation may have a variety of influences on ultimate economic goals: the quantity and diversity of consumption, employment, income distribution, and so on. To the extent that innovation has a beneficial or detrimental impact upon such objectives, these influences will be important for policy formulation. This issue is taken up again in the last section.

To sum up, it is possible to argue that the market may generate suboptimal quantities relative to other forms of investment; discrimination against basic research and the innovations of new and/or small firms; a sub-optimal allocation of research resources across the set of potential opportunities; and socially mis-timed innovation. The list is not exhaustive but it suggests a basic case for government concern. Policies might be, directed towards stimulating R & D overall, directing it in particular ways and attempting to find optima in the timing, sequence and mix of innovation projects.

As must therefore be apparent, there is ample reason to suspect that the market may not be able to generate a welfare optimum in either the quantity, type or mix of R & D and consequent innovation. Two rather general types of argument need to be considered, however, which suggest that this is by no means a case for wholesale government intervention. The

18. Kemp (1980).
19. Kamien & Schwartz (1982), esp. Chs. 4 and 5, report the debate and give extensive references.

first line of argument suggests that governments may not actually be able to do any better than the market. The second asserts that most innovation occurs in response to demonstrated market demand and that, by implication therefore, it cannot be seriously sub-optimal.

To illustrate the first argument, recall the discussion earlier relating to uncertainty. Given that no individual firm knows what all other firms are planning to invest (in physical capital and R & D), no individual firm can know the future course of prices for the goods which it produces (and will in future produce) since those prices will depend upon how much and what type of investment other firms do. The individual firm can no longer be characterized as a price-taker, but given that each firm must know future prices in order to invest efficiently, this means the price mechanism has broken down. Can government help? It has been argued[20] that centralized decision-making is the way out; but any aggregate investment plan would require a concomitant rate of saving by households, and without centralized information about individual consumer preferences, government could not hope to generate such savings in an optimal way[21]. Even with such information it might be added, a government would have to resolve difficulties of the kind raised by Arrow's Impossibility Theorem[22] on social welfare functions and find instruments to shift consumption which, in themselves, were non-distortionary. It is recognized, of course, that such instruments are not easy to devise. Not all problems are as intractable as this, however, and as will be seen in a moment, the fact that we live in second or third best world may even offer its own solace.

As a matter of logic, priority should usually be given to rooting out the cause(s) of divergence(s), in preference to imposing further distortions designed to offset existing ones. For example, finding means of improving information flows about new technology use attacks the roots of uncertainty where a subsidy on research may lead only to wasteful duplication of effort. On the other hand, it is readily agreed that some distortions are unavoidable. R & D expenditure undertaken by one firm generates information from which other firms might (in due course) benefit, and so generates either a technological or pecuniary externality. With or without patent protection, innovations can often expect to enjoy a temporary monopoly while potential imitators unravel the technical secrets of a new process, acquire necessary know-how to implement it and gather enough experience to learn by doing.

20. Graaff (1951), p. 105.
21 Phelps (1965), Ch. 4.
22. Arrow (1963).

In cases such as these, there would seem to be a need to appeal to the Theory of the Second Best[23].

In a second best world, the Pareto assumptions are violated and distortions are reflected in the failure of the market to generate the marginal equalities which characterize a first best optimum. In the case of a monopoly good, for example, price exceeds marginal cost whereas in another part of the economy competitive conditions may ensure that price for a different good equals marginal cost. Utility maximizing consumers operate where their marginal rate of substitution between the two goods equals the ratio of their prices. In this case the price ratio is not equal to the ratio of marginal costs, i.e. the marginal rate of transformation between the goods, and so mrs≠mrt. For policy makers, the significance of the Theory of the Second Best is that it derives conditions which characterize an optimum, *given* the distortions, and to the extent that government is involved in productive activity, implies appropriate action. In the case of private sector R & D activity, we have already condsidered the argument that because firms are able to capture for themselves only a fraction of the social benefit their efforts generate, there may be a socially sub-optimal supply of inventions. Here, the Theory of the Second Best implies that a government could engineer a second best optimum by pricing below marginal cost public sector outputs which were complementary with R & D effort, thus encouraging demand for such goods and, indirectly, increasing the demand for goods embodying the new techniques.

For policy purposes there are, however, two extremely serious problems involved in appealing to second best rules. First, to apply the rules to even one pair of goods it is necessary to know the ratios of marginal costs and marginal rates of substitution, the degree of complementarity and substitutability between the goods and the effects upon the marginal costs of producing one good and changing output levels in the other. Even where simple manufactured goods are concerned, this presents a significant agenda in applied research for policy-makers, but given the uncertainty surrounding the determinants of research activity and its success rate, the problem becomes particularly acute. Second, a second best optimum can be guaranteed only if the second best conditions are met for each and every pair of goods and services. Unless this is the case, there can be no guarantee that intervention — even if successful in all but a few cases — will have achieved an improvement with respect to the initial situation. The informational and computational burdens implied by this result are almost beyond imagining. The

23. Lipsey and Lancaster (1956).

wide-ranging and unpredictable influences of R & D only serve to intensify the problem.

Considerations of this kind suggest that almost no intervention could ever be justified in terms of predicted or predictable welfare improvements. But it has been argued[24] that because, on the contrary, we live in a world which fails even to be second best, relatively straightforward policy prescriptions might, after all, be in order. The information required to construct the second best conditions or rules is often not readily available and usually costly to obtain. It is also costly to apply the rules. These costs, together with the distortions relating to a first best optimum, constitute the characteristics of the third-best world in which, in fact, we live. The rules which it is appropriate to apply in such a world depend upon what information policy-makers have and what attitudes to risk they possess.

There is informational *poverty*, if knowing the amount of information they will acquire, policy-makers find that information is insufficient to let them form a reasonable probabilistic judgement about: (a) the direction and extent of divergence of the second best optimum from that which would result from applying the first best rule in the presence of the second best distortion; and (b) the shape and skewness of the concave function relating values of the objective function to degrees of divergence from the first best rule[25]. (This function reaches its maximum, of course, when mrs = mrt for all pairs of goods, and must take lower values whenever mrs ≠ mrt for any pair.) Assuming non-risk-loving behaviour by policy makers it is easily shown that the optimal policy is that implied by the first best rules. Second best considerations relating to the overall matrix of effects should not discourage the use of appropriate fiscal measures whenever the direction is known i.e. subsidies for an external benefit, tax for an external cost[26]. Not to act in cases like this is to ignore information which is specific and useful. In the case of R & D, of course, this may be of little comfort. Innovation generally imposes external diseconomies as well as benefits so that this "escape route" may not, after all, be available unless it is fairly clear which way the balance will tilt. Nonetheless, assuming that at least this much is known, third-best theory suggests that "piecemeal policies can be justified"[27] and the agnosticism of second best theorists put to one side. Given the piecemeal nature of so much policy for technology, this is perhaps nice to know.

24. Ng (1979), Ch. 9.
25. Ng (op. cit.), p. 231.
26. Ng (op. cit.), p. 235
27. Ng (op. cit.), p. 235.

The wide-ranging but often unpredictable and complex effects of techno-logical change may well lead many to believe that policy-making in the area must indeed be beset by information poverty. In some cases, however, there may be enough information available at least to permit a reasonable probabilistic judgement about points (a) and (b) above. If so, there is said to be information *scarcity* rather than poverty. In such cases, however, there still is not enough information to permit the derivation of rules ensuring a second best optimum. Third best theory suggests that policy-makers examine the information and decide whether, in light of it, there are reasonable grounds for amending the first best rule appropriately if there is fairly precise information about the way in which an external (dis)economy generated in producing good X operates via the substitutes and complements for X in creating further distortions — even though all the economy-wide ramifications are not known. Similarly, if policy-makers had the information to assure them that such (dis)economies would *not* create any further distortions at the first round (again, in ignorance of other indirect effects), then applying the first best rule would be optimal.

While this discussion suggests that, after all, there is a place for policy in pursuit of enhanced welfare, it leaves unresolved what might be termed a meta-problem. This concerns the distinction made between the two "levels" of information. Whether there is informational poverty or scarcity hangs on "a reasonable probabilistic judgement"[28] — which can be made only by policy makers and their advisers — about the sufficiency of information in relation to points (a) and (b) above. Yet surely there is room for intelligent men and women to disagree about whether they have sufficient information to form such a judgement or not. In the case of information about the impact of R & D, the scope for disagreement is manifestly substantial — and hence the appropriate strategy to adopt is also likely to be a matter of dispute.

The second argument against intervention is that such innovation as occurs does so in response to market demand. (This is the sort of stand epitomised in the work of Myers and Marquis, (1973) for example). If this is accepted, it is a short step to arguing that innovation cannot be seriously at variance with social preferences. The process of innovation, considered at greater length in the next section, has been shown, however, neither to flow purely from market to producer (need-pull) nor purely from laboratory to market (technology-push). Both elements are involved in a complex evolutionary process. In any case, it ought to be clear that consumers or users simply cannot always conceive of all the innovatory products which might, in prin-

28. Ng (op. cit.), p. 231.

ciple, become available. Whether and where governments might successfully intervene to clarify demand and overcome imperfections in information
are questions which cannot at this point be answered. But they must be borne
in mind.

Where to intervene?

If it is accepted that there may be some role for policy, a second question
relates to how and where to intervene in the innovation process. This calls
for a brief review of what is known about the process.

Given the apparently chaotic and impenetrable mystery of technological
change, the taxonomy of invention, innovation and diffusion has proved a
popular framework for debate. As a means of putting structure on the overall
process of technological change, this taxonomy is an admirable device. A
priori, it seems likely that the factors which influence the generation of inventive ideas will be significantly different from those associated with the
conversion of those ideas into a marketable product, and different again
from those which determine how quickly, if at all, a new process or product
will spread through the community of potential users. Unfortunately much
research and policy-making effort has proceeded on the assumption that
the taxonomy alone necessarily implies a linear dynamic process over time.
This so-called *linear model* suggests that all technological change starts, in
time, with scientific discovery which, in turn, gives rise to an intervention
which, subsequently, is developed to the point where it can be marketed as an
innovation which, later still, diffuses in use or consumption through the
economy. Notice that the time-dependent and unidirectional nature of the
last sentence is in marked contrast to the classificatory nature of the pure
taxonomy (which would retain the essence of its value, whatever the order in
which the three elements were considered).

Now, it cannot be denied that some important technological change does
occur linearly: the atomic bomb has been cited as a clasic example of "an
innovation that resulted entirely from the scientific discoveries of men and
women working in universities, whose motivation was the advancement of
knowledge, not practical utility"[29]. But the unidirectional, linear model ignores
the abundant evidence that much technological change occurs in response
to perceived market needs[30]. Myers and Marquis (1973) went on to explore
this theme further, concluding that 75 per cent of the 657 innovations they

29. Ronayne (1984), pp. 44-47.
30. Schmookler (1966).

examined in five different industries "could be classed as responses to demand recognitions" (p. 31). Other studies have reached similar conclusions, and the single most important factor determining *successful* innovation certainly seems to be a close monitoring of potential markets[31].

While they are clearly far too important to ignore, it is as dangerous to see demand-side influences as the sole prompt to technological change as it is to consider only the linear view and some have argued that business-men take the existing, inherited range of productive processes as their "ine-vitable" starting point. Almost all processes have constraints and short-comings and these "create internal compulsions and pressures which, in turn, initiate exploratory activity in particular directions". "Compulsive sequen-ces" start as efforts are made to resolve imbalances in the existing network of technological interrelationship. They continue as an initial change pro-vides the stimulus for change elsewhere[32]. And in relation to new production techniques it has been argued persuasively that, in the petro-chemical industry, post-innovation learning and incremental improvements are three times more important in reducing costs than is the introduction of the innovation itself[33]. We have here two important elements driving the overall process of technolo-gical change, yet neither is in any direct or obvious way the consequence of demand side pressure.

Even a reasonably accurate description (let alone analysis) of technolo-gical progress is bound, therefore, to be rather complex. At the least, *both* "technology-push" (essentially supply-side) and "demand-pull" factors need to be considered — and preferably interactively. Such a view flows quite na-turally from the argument[34] that market demand factors not only do not dominate all others, but could neither be expected nor shown to dominate. Better, a framework is needed which shows all of the important stimuli for change. In some cases major change does follow the linear model; in others, it arises from a market need which has been communicated or is obvious to researchers. Very often change proceeds incrementally. A new product or process is tested in use or in the market and feed-back mechanisms are set in motion which lead to further developments and improvements. Some-times the changes are the work of the innovator; sometimes innovators may find themselves hurried from the market by a fast-working imitator. Further, all of the forward and backward linkages throughout the input-output re-

31. Science Policy Research Unit, Project SAPPHO (1972), Meadows (1968) and Thomas (1971).
32. Rosenberg (1976), pp. 110-111.
33. Enos (1958), p. 180.
34. Mowery & Rosenberg (1979).

lationships in the economy imply a continuous flow of ripple effects and repercussions between any given activity and a myriad of others. Some of this complexity can be captured by focusing analysis on what has been termed the *productive segment*[35]. This unit includes the physical process, product, and the characteristics of inputs and product demand which are related to the product. Clearly, a vertical span of some general kind is at the core of this notion, and it has been shown that *product* innovations at earlier processing stages have received their main stimulus from actual or anticipated opportunities for *process* at higher process levels. At the same time any product innovation may itself stimulate process change at higher levels[36].

With so much going on in the process of technological change, two things at once become clear. Policy built on a simple linear perspective is almost bound to be doomed, not only because it can hardly be flexible enough to serve, but also because it will appear *simplistic*, if not irrelevant to those actively engaged in the business of generating change. Second, whatever policy is considered, it must have the potential for many different guises and operating in a host of ways on a large range of variables[37].

If a government is interested either in stimulating or shaping innovation, therefore, no simple nostrums are available to it. To the extent that the process does start with basic research, there may be reason to support it — but it will be at least as important to ensure that the findings of scientific research (wherever they originated) are made easily and widely available to all potential users. Whether the perspective taken is that of demand-pull or compulsive sequences, we may generally expect to find more problems in search of solution that *vice versa*. Some solutions will require appeal to recent scientific advances; others may be found in the application of practical engineering. Confronted with problem-solving activity, governments may be concerned to minimize wasteful duplication of R & D effort; they may wish to ensure that the framework of the incentives is appropriate to encourage the fairly rapid discovery of solutions; they may seek to stimulate parallel research plans; they may try to oil the wheels of technology transfer — internationally or among firms. We turn now to a brief and selective review of policy tools which have been tried.

35. Abernathy & Townsend (1975).
36. Abernathy & Townsend (op. cit.).
37. Freeman (1980), p. 321.

Tools for intervention

A very wide range of strategies has been applied in attempting to stimulate and/or direct innovation. The history of such policy is outlined in Ronayne (1984) and extensive discussion of policies and country experience can be found there, and in Pavitt and Walker (1966), Johnston and Gummet (1979), Tisdell (1981) and Ronayne. Policy tools which are available include direct government participation in research; attempts to stimulate private research by placing government contracts with innovators; all manner of subsidies, tax reliefs, loans, and investment allowances; centralized co-ordination of research activity; the patent system; attempts to reduce market imperfections; honours and awards; general economic management aimed at providing the most attractive climate for innovation; and educational and training schemes.

Direct government participation is perhaps most easily justified in relation to basic research. Pure research often has all the characteristics which will tend to lead the market to underinvest in it: large investment costs, uncertainty of outcome (with respect to the outcome itself, the date of solution and eventual value), and potential inappropriability of rewards. The other leg of basic research, strategic research, and, in addition, significant elements of non-basic research and development are often justified in terms of national defence requirements. General arguments in support of some unspecified amount of expenditure do not, however, provide much assistance in practical policy formulation. Here, the essential questions relate to *how much* to spend in total, in relation to other public investment, and in each category? Clear answers to such questions are never easy to find. The economist's instinct is to look for social costs and benefits as the basis for determining a social rate of return, but since the essence of basic research is extending from the known into the unknown (or, at best, only suspected or hypothesised), forecasting returns must, by its very nature, embody extremes of uncertainty. In many cases, moreover, the uncertainty is of the most intractable form, relating to events for which there is little or no precedent, thus rendering redundant any probabilistic analysis based upon the premiss of repeated experiment. The uncertainty here relates both to the potential outcomes of the research and their applications. Traditionally, scientists have often wanted to argue[38] that they are the people best equipped to determine the type (and, by implication, quantity) of basic research undertaken. There is clearly an irreconcileable tension here. Government's efforts to direct science may

38. See Ronayne (op. cit.) for the views of Polanyi (p. 78) and Baker (p. 82).

arrest its progress: but if science requires public support for expensive research, it cannot expect Governments to absolve it from the responsibility of putting resources to best use. Governments may find it self-defeating to guide science but equally science may be poorly situated to appreciate or foresee the wider socio-economic implications of its efforts. Governments may have little better idea of these implications than science, but they are charged with acting responsibly. This may suggest a minimal allocation of resources to "science as high culture"[39] but an iron resolve to try to find a sensible way of ranking projects. No ranking procedure can ever be more than relativistic and indicative; but the *process* of ranking should at least contribute to exposing relevant information and thereby improving decision-making.

The last point made relates to *improved information flows*. Particularly with the diffusion of information technology there would seem to be scope for setting up databases containing a carefully organised catalogue of advances, and of market opportunities for particular types of innovation. Such a strategy has potential, however, only to the extent that such databases can be widely and easily accessed, and that firms' managers are motivated and trained to search effectively. This is relevant to government support of R & D itself since more than 45% of knowledge inputs into firms has been found to derive from government-funded technological institutes and universities[40]. Transferring knowledge from source to user is clearly a significant issue, therefore. Educating potential users is another.

It has been observed that government's immense *purchasing power* has often been used as a blunt instrument but "could with benefit be finely honed"[41] While this approach might in principle either stimulate or direct innovation, it also has its disadvantages. For example, products specifically tailored to a government buyer's needs may reduce subsequent export prospects and resulting low-volume production may inhibit the growth of profits required for future investment in R & D.

Given the uncertainty and potential waste associated with decentralized research planning, another strategy might be to *co-ordinate* firms' activities or even to exercise centralized control. This presupposes that governments can always do better at predicting the future than decentralized decision-makers. It also ignores the reality that industry research associations are not only rational (for reducing uncertainty) but have also come spontaneously into existence.

39. Johnson, H.G. cited in Ronayne (op. cit.), p. 75.
40. Pavitt & Walker (op. cit.), pp. 22-3.
41. Watkins & Rubenstein (1979), p. 197.

The historic response to the welfare problems raised by innovation was, and still is, the *patent system*. Patents provide a vehicle whereby inventors can gain a means of appropriating some of the benefits which their ideas generates for society. This should stimulate inventive activity. It should also act to discourage the inventor from turning businessman in an attempt to embody his idea in a commercial product when existing business would do the job more efficiently. On the other hand, to the extent that the patent system imposes royalty costs upon the user, it may reduce usage of the invention to a suboptimal level. It should be noted that patents cannot do anything to reduce the *technological* uncertainty associated with R & D as a process. They merely provide greater certainty that some of the benefits generated by an invention will flow to the inventor.

The patent system has always been recognised as an awkward compromise, and if innovation is very fast-moving or imitation costly and slow, patents may become irrelevant[42]. On the other hand, complete protection from imitation can only rarely be guaranteed and "inventing around" patents is an important competitive strategy. It is also now recognised that *optimal patents* will vary from industry to industry and from one invention to another. The relevant calculations require information on R & D expenditure and the associated probabilities of successfully completing a research programme by specified future dates[43]. It is unclear how much of this information could, even in principle, be known. Other recent work suggests that insufficient attention has been paid in the past to *prizes* and *contractual research*, compared with which the range of situations in which a practical patent system dominates may be narrower than commonly believed[44].

Subsidies and other *fiscal incentives* have always been the first resort of welfare economists seeking to correct divergences from welfare optima. In this case, however, several problems are apparent. First, should *all* forms of innovating activity be made the subject of incentives? Some innovation involves developing new processes to make wholly new products, but much of it comprises a gradual evolutionary process : slight modifications to the product in response to market feed-back, gradual improvements in processing as labour force and managements learn to use and adapt new techniques[45]. Dividing up innovatory activity into that which *needs* incentive and that which would have occurred anyway must in a sense be arbitrary therefore. Second,

42. Mansfield et al (1981), p. 916.
43. Dasgupta & Stiglitz (1980).
44. Wright (1983).
45. Rosenberg (1976), Ch. 6, Enos (op. cit.).

incentive schemes share with patents an incapacity to reduce technological uncertainty. By reducing expected costs, subsidies accompanying investment can increase the mean of the distribution of anticipated returns, though will not affect its dispersion. This should be an inducement for risk averse decision-makers but in turn raises another problem, that of moral hazard. To overcome that difficulty, it may argued that a more effective policy would be to offer *loans* instead, with an expectation of recovering the government contribution from successful R & D[46].

Whichever approach is taken, governments will almost certainly find themselves confronted with a selection problem: which innovations, firms and industries are most deserving of support? It has often been argued that governments are rather poor at "picking winners", but that claim needs to be interpreted with care. Even professional R & D managers are not good at picking winners. It was once found in the USA that for every 100 projects begun, only 31 ever reached the market and of those only 12 were successful[47]. Governments can expect to do better than this only if they know something about future markets which firms do not — in which case it would seem most cost-effective simply to pass on the information. On the other hand, there is a difference between projects and industries. In the climate of the 1980s, it should be possible to predict that "microelectronics or biotechnology will have greater growth prospects than sausage casings or rain coats..."[48]. That granted, the difficulty of defining the most appropriate policy response still remains, however.

Given the problems of selection, governments have inclined more recently to seek *ways in which the market might be made to work better.* Small firms in particular may be discouraged from innovating when they note the absence of potential lenders in the capital market and/or the high learning and transaction costs of trading in knowledge markets. In response to the first problem, governments may make loan finance available directly, or might set up new lending institutions with suitable structures of incentives to attract funds from the existing capital market. The second problem arises from the fact that either as a licensor or licensee, a firm wishing to trade in new technological knowledge initially faces costs related to learning about participation in this market and subsequently may have to commit resources to search and to negotiating contracts. Trading is therefore not the costless, instantaneous process theoretically underlying the achievement of a P-O,

46. Gannicott (1980), p. 300.
47. Mansfield et al (1971).
48. Barry Jones, Minister for Science and Technology, Australia, quoted in *Australian Financial Review*, May 1984.

via the competitive mechanism and, proportionally, the effect of such costs on small firms is relatively great[49]. A system of government extension offices or subsidised negotiation might be solutions here.

Finally, the single most powerful determinant of commercial success for an innovation seems to be attention to, or accurate estimation of potential markets (Pavitt and Walker 1976, p. 27). In this connection, it is of interest to note the activity of the Greater London Council through its Enterprise Board. One move which it has made is to seek out consumers who have expressed dissatisfaction with existing goods and services and ask them what they would prefer instead. Fragmented voices of discontent can thus be articulated to indicate needs and act as a positive guide to subsequent development. A special attraction of this approach is that it enables producers to draw upon the dynamics of evolving preferences. In just the same way as existing production technology evolves automatically as imbalances are resolved, only to reveal new ones, so with consumption technology, dissatisfactions arise with the existing structure. only to reveal new opportunities. The potential for market failure exists here because producers may perceive sustained demand at existing prices as a mark of approval (when in fact consumers continue to buy, *faute de mieux*, and cannot signal their preference for something else, simply because it does not exist).Even if complaints are voiced, they may then tend to be ignored. It is a mechanism to reveal what consumers would prefer which is needed[50].

As might have been hoped, therefore, the range of policy tools is indeed wide and varied. Effectiveness, however, is a different matter.

The impact of policy

Throughout the discussion so far, the unspoken assumption has been made that policy initiatives can actually achieve a measurable effect, irrespective of whether for better or worse. In this section, some of the evidence on this point is reported and its implications discussed.

One particularly interesting study in this respect[51] sampled 15 firms in Britain, the organizations varying widely in size, location and productive activity. The comments elicited are too distinctly negative to be able to ignore. In brief the study reports: (i) most government incentives and support programmes for R & D and innovation have no effect on most firms; (ii) aware-

49. Lowe (1984).
50. Greater London Enterprise Board (1983).
51. Watkins & Rubenstein (op. cit.).

ness of many schemes was low by any standards; (iii) the generally low level of interest in government assistance reflected in claims by most firms that they would prefer complete government disengagement from industrial involvement, even at the expense of subsidies and technical support services; (iv) government purchases of innovative products certainly did have an impact, but sometimes with unfortunate results; (v) grant-aided industrial research associations controlled by subscribing members with related needs were perceived as declining in value − except in relation to their capacity to provide technical information to smaller firms which would otherwise be denied access to it; (vi) with only a few, if important exceptions, financial inducements to modify decision-making behaviour generally failed to have that effect, and to the extent that they were made were regarded as bonuses justifying decisions already taken, and (vii) incentives were often concentrated too specifically at the R & D end of the process of innovation.

Further evidence in similar vein may be gleaned from studies of the economic impact of patents. Australian surveys[52] suggest that the patent incentive is apparently not an important determinant of the level of overall measured industrial R & D activity. Similar conclusions apply to the UK and Canada[53]. More pointedly still, a senior corporate vice-president is reported to have said: "In the electronics industry, patents are of no value whatsoever in spurring research and development"[54]. And several studies have made the more general observation that the patent system is too slow to be of use to businessmen operating in the fast-changing, intensely competitive world of high technology centred, at present, on semiconductor electronics and biotechnology[55]. A general exception to this agnosticism seems to lie in pharmaceuticals. In the UK for example the pharmaceutical industry had a higher proportion of production based on patents (68%) than any other industry[56].

The evidence cited thus far tends in the main to suggest that existing policies may have had only insignificant effects, and many would view this as conclusive proof of the case that government should steer clear of the whole area. But there is another side to the picture. First, some policies have offered clear benefits which probably would not have otherwise accrued; second, there may have been implementation problems with existing policies, while potentially useful alternatives may not have been put to the test; and

52. Mandeville et al (1982).
53. Taylor & Silberston (1973) and Firestone (1971).
54. Mandeville et al (op. cit.), p. 108.
55. Mandeville et al (op. cit.), pp. 107-109.
56. Taylor & Silberston (op. cit.), pp. 201-208.

third, it is always hard to know whether, in aggregate, innovative activity would have been greater or less in the *absence* of relevant policies.

On the first point, there are at least some innovating firms in existence in UK which would have gone out of business had it not been for the National Research Development Corporation[57]. The NRDC has employed a number of mechanisms in supporting development projects in industry but usually shares both their rewards and their risks. Again, legislation in respect of technical standards can, for companies subject to these rules, shape and direct large parts of the innovation effort. Such constraints can be productive, indeed, not only by ruling out options perceived to be socially detrimental but by forcing firms to think of better ways of solving problems. In a very real sense, necessity here can be the mother of invention. As a final example, universities and other government-funded research institutions continue to generate a steady flow of major and minor technological advances. Some, even many, may not be commercially viable, but others undoubtedly are, and the burden of proof is on the doubters to show that at least as many "winners" would have been turned up by the market, and at smaller cost, before closing the argument against funding of this kind.

On the second point, sheer ignorance about the existence of policies is clearly a frequent problem. On the third, innovative activity is subject to many and, over time, changing stimuli so that isolating the influence of policy from all others under *ceteris paribus* conditions is an almost impossible task.

Broadly speaking, science and technology policy could probably be characterized as being concerned to influence the more efficient working of markets and overcoming the incapacity of the market mechanism to deal efficiently (if at all) with certain phenomena associated with innovatory change. This is essentially policy making at the micro level. The principal macroeconomic objective to which such policy is likely to be directed is, it seems clear, growth. Two ultimately related points now need to be made. First, growth built on innovation can be achieved only to the accompaniment of changes in business organisation, industry structure, social arrangements and other parts of the technology. At the firm level it has been observed that "the ability of the organisation to *adapt to* its new technology seems to be a crucial factor in innovation, and one that is often ignored"[58]. As for structural change, many students of growth have recognised that as economies expand, so individual industries emerge, prosper and decline. The reason for

57. Watkins & Rubenstein (op. cit.), p. 196.
58. Macdonald & Lamberton (1981), p. 11 (Italics added).

this should be apparent. One of the main thrusts of technological innovation is to raise productivity, but simultaneously this raises per capita incomes. As real incomes rise, so patterns of consumer expenditure change, for as empirical studies widely demonstrate, the proportion of income spent on any type of good changes as per capita income rises and demands for new types of product also emerge. Since industrial structure is related intimately to the structure of demand, it is entirely to be expected that different industrial sectors will have different and non-proportional growth rates both from each other and from that of aggregate output. This means that most modern growth models — which assume equiproportional growth rates in demand — have to be considered unacceptable[59]. It also means that growth-oriented policy cannot but take account of the fact and implications of structural change. As for changes in social arrangements, it should be sufficient to draw attention to the way long-distance commuting has accompanied the widespread diffusion of the motor car, and to the ways in which work stations in the home may revolutionise the notion of the "job".

The place of innovation policy

Thus far this introduction has been devoted to discussing in rather general terms the ways in which policy might have a role to play in either correcting distortions or helping the market to perform better in achieving welfare. Suppose, for better or worse, governments do choose to involve themselves in this area (and most have), where does policy in this area fit into the usual framework adopted for policy analysis?

It is conventional to list the principal objectives of government policy as full employment, a stable price level, a "satisfactory" growth rate, balance of payments stability and income distributional equity. In a closed economy, the efficient allocation of resources during each period and over all periods should ensure that the growth of output target is met. The implications for employment are more ambiguous but (see below) may not always be as troublesome as is popularly believed. (This issue is taken up in a later essay)[60].

Distributional factors are intimately bound up with technological choice, relative factor shares depending upon bias in technological change and the personal distribution being influenced by employment opportunities and attitudes towards risk. On the other hand, there are clear dangers in *using* technology policy to influence distribution, mainly because maximum bene-

59. Pasinetti (1981), pp. 68-70.
60. Williams, Ch. 4.

fits from innovation are to be reaped by allowing a maximum of spontenaity on the supply side and flexible response to market requirements from the demand side. Other instruments have long been available for adjusting the income distribution generated in the market. There are also relationships between technology, the price level and inflation. The faster is technological progress, the faster factor productivity should rise and the less "cost push" should exert upward pressure on the price level. On the other hand, if labour feels it is receiving an insufficient share of the benefits derived from higher productivity, trade union agitation may be expected and inflationary pressures may re-appear from a different angle. Lastly, the external balance of trading economies may be affected in a number of ways by technological progress. Countries which are slow to innovate will find it hard to compete in world markets with countries which are more successful innovators, and countries which innovate fast may also enjoy net gains in patent and licencing fees (see the essay in this volume by Silberston and Ledic). These points are now discussed in turn at greater length.

The second point to be made here is that, given the far-reaching and complex effects of technological innovation, governments will find themselves having to confront the question of how other goals will fare as innovation-driven growth proceeds. The level of employment will be one central consideration. At the level of individual productive activity, productivity-raising technical progress must, by definition, result in the reduction of per unit employment of some or all inputs, one of which may (but need not) be labour. While the impact or first round effect of productivity-raising innovation will often entail labour displacement, subsequent effects may well lead to re-employment and net employment creation. Such positive effects include the fall in the relative price of the output of the firm or industry experiencing technical change and the rise in real incomes accompanying the productivity rise. In fact, it has been shown that it is often those industries whose labour-productivity rises most quickly which also experience most rapid employment growth[61,62]. The factors which determine whether an innovation will create or destroy employment, once its effects have worked through, include the form of the technical progress itself, technical substitution possibilities between inputs in the innovating sector, wage flexibility, aggregate demand effects and the substitutability in consumption between goods produced in the innovating and non-innovating sectors[63]. Perhaps not surprisingly, there is the

61. Salter (1966).
62. Metcalfe & Hall (1983).
63. Sinclair (1980), Neary (1981), Hall & Heffernan (1985 forthcoming).

potential *a priori* of almost any outcome – including net employment creation in the wake of labour-saving innovation. In the absence of highly flexible
wages and prices, the ambiguity of potential outcomes also appears in a dynamic multi-sectoral framework. Here, productivity raising technological
progress combined with the approach to saturation of demand for innovation
creates the basis for new employment by creating new products. In fact, policy-
makers interested in employment objectives are faced with an inescapable
choice, given even a modicum of price rigidity. Either they contribute to
maintaining full employment by ensuring that new products are created
fast enough to offset the negative effects for employment of the first two
factors; or, failing that, they must live with a situation in which more and
more leisure has to be taken by the workforce. If the full employment target
is perceived in terms of fixed quantities of work and leisure hours per period,
this gives a new urgency to the task of demand management[64].

The debate over links between technological changes and employment
suggests several areas for constructive government interest. The management
of aggregate effective demand has implications both for maintaining the
flow of product innovations and for smoothing the adjustment from episodes
of initial labour displacement through the period of subsequent employment
regeneration. Further, training schemes can assist on the supply side to
fit labour for new tasks associated with producing new goods and services.
Whether the employment objective *should* be viewed in terms of a fixed
quantum is, however, the more fundamental issue. The focus of attention in
this area of policy making must increasingly be the question of how much
leisure and work society optimally wants for itself, a question which is likely
to render redundant historical estimates of what "full employment" means
in quantitative terms.

The structural charges inherent in the innovating economy also have
implications for the price level objective. It is clearly implausible that an
economy should be in a perpetual state of structural adjustment without,
at the same time, observable change with respect to the structure of relative prices. To understand this, suppose for a moment that only one firm
in one industry is a cost-reducing innovator. Cost reductions will enable the
firm either to make increased profits at given prices or reduce price to increase its market share. Both the inducement of profit and the threat of
lost market power will act as a spur to imitation or independent innovation
by competitors and such competition can serve only to drive prices down.
Thus, the price of the output of the innovating industry will fall relative
to that of other industries' output. More generally, it can be argued that

64. Pasinetti (op. cit.), Ch. V.

those industries whose productivity growth is most rapid are also likely to be those whose relative output prices fall most[65]. This (over-) simple description of the market at work also has a normative parallel. For any given pattern of demand, the relative price of the output of a cost reducing industry should fall, otherwise the price of the output of a cost-reducing industry will give a distorted reflection of the value of resources employed in this activity. How does this bear on the goal of price stability?

Suppose price stability means a stable value for the average price level. This goal is compatible with that of permitting real wages to rise at the average rate of productivity increase so long as the overall profit rate remains constant over time and the nominal wage, on average, also rises at the average productivity rate. But keeping the average price level constant requires firms and industries which achieve above-average increases in productivity also to cut their prices by an above-average amount. Unless they do, real wages (nominal wages divided by average price level) will rise by less than the average rate of productivity. It may well be argued that, at least for some periods, this is quite desirable – since a fall in the real wage implies a rise in the profit rate, and higher actual and expected profits may be the most effective mechanism for raising investment in innovation. But if governments have promised or trade unions think their members should simultaneously receive the full benefits of any productivity increase, then there must be pressure to raise money wages at a rate faster than the average rate of productivity increase. This in turn will lead to a rise in the general price level.

This analysis brings to light an uncomfortable conflict in the area of policy making for innovation. At the micro level we have seen that there may in some instances be a persuasive case for patents – and this case may be strengthened if it can be shown that the patent system increases the general rate of innovation and thus helps in achieving the employment goal by encouraging a faster introduction rate of new products. On the other hand, patents inhibit the forces of competition in driving down the price of goods produced with innovatory processes and hence may be a contributory factor to inflation. A more general issue is the distributional one : irrespective of the implications for the price level, *should* labour see its real wage grow at the average rate of productivity growth, or at some other rate which implies either a rising or falling profit rate?

The conventional approach to the question of income shares has been to take the technology as given. With a given technology, the elasticity of

65. Salter (op. cit.) and Metcalfe & Hall (op. cit.) offer empirical confirmation.

substitution between inputs in production is exogenously determined and it can be shown that relative shares depend on the value which it takes. In brief, if the elasticity of substitution is great than, equal to, or less than unity, a rise in the ratio of profit rate to labour's wage will lead, respectively, to a fall, no change, or rise in the share of capital relative to labour.

Several points should be borne in mind about this. First, even though the *relative* share of a factor (say labour) might be falling, the *absolute* value of labour incomes may well be rising — even though, by definition, more slowly than incomes to capital. If technological progress is biased so as to diminish labour's relative shares, or to drive it in labour's favour, the penalty might be to slow down the investment rate as profits grow less quickly (or not at all) and as a consequence penalise growth. This in turn would lead to a slowing down in a growth of overall incomes and could mean, eventually, that the absolute value of labour incomes might turn out to be lower than they could have been under a regime of lower relative labour share but higher investment. It has been argued[66] that from an evolutionary point of view variables such as the elasticity of substitution should not be regarded as exogenous data but as potentially manipulable by government policy. While this may be a sensible view in relation to finding substitutes for scarce inputs (such as energy), in the case of income distributional policy the argument would be harder to sustain. Why not use taxes and subsidies instead?

Finally, notice that the relative shares approach takes a functional view of the income distribution whereas the personal distribution may often be of more pressing concern. It is quite inevitable, we must suppose, that an economy undergoing innovation-driven structural change will also have a marked variance in and probably skewed income distribution. As new industries come to prominence, returns on both the human and physical capital required in those industries may be expected to be relatively high. In the case of human capital, skills required in the new industry will be scarce precisely because the industry is new and its specialised requirements will not be immediately available from the existing pool. Labour with such skills will be able to enjoy a quasi-rent. In due course, the prospect of relatively high incomes will induce labour to acquire the skills required by the new industry — and this will have the effect of eroding existing quasi-rents. While this suggests a tendency to reduced variance in the distribution over time, the trend is likely to be offset by the emergence of further new industries with their own new demands, generating their own higher-than-average incomes. Again, the observation that R & D and innovation activity is spread une-

66. Nelson & Winter (op. cit.), p. 186.

venly across firms suggests a variety of business attitudes towards risk. If the business community is viewed, schematically, as divided between the risk-loving and the risk-averse, then the distribution of incomes within the first class is likely to be much more dispersed than for the second, but with a higher mean. The summation of the two distributional generates a skewed aggregate distribution[67]. In each case innovation is intimately associated with income dispersion, but in neither is it clear that policy-makers should be concerned. On the contrary, efforts to tax away quasi-rents from those with scarce skills or the fruits of risky investment from entrepreneurs will only serve to slow necessary structural change and reduce the incentive to innovate.

As for external balance, it is fairly easy to make a case to the effect that countries which innovate most successfully will also enjoy the benefits of a healthy export market, both in innovative goods and in knowledge itself, if tied up in licensing agreements, turn-key arrangements and so on. To a significant extent, policies aimed at raising or maintaining the domestic growth rate by stimulating innovation would seem to be consistent with the aim of maintaining external balance. In principle, an innovating economy should be a strong competitor in international markets. At the same time, a rapidly growing export market should sustain a healthy growth rate. But on closer examination, the appropriate course of action may not be alto-gether easy to define.

In general, it makes good economic sense for countries to specialise in areas of production in which they enjoy comparative advantage in international trade. The same principle should also apply to R & D — but it is not at all clear where any given nation's areas of comparative advantage in R & D reside. It is tempting to believe that such areas will conform closely to a nation's areas of comparative advantage in traded goods and services, and to some extent this probably will be the case. If a country has built upon its comparative advantage in trade to build up specialised industries in those areas, it should also have a specialised workforce which has learned (and is still learning) by doing in those areas. On the other hand, this will not se-cure comparative advantage for the country in R & D if other countries have learned more in these areas, relative to other types of activity than the country in question. Rapid learning about and development of *other* countries' technology has, in the view of many, been the hall-mark of recent Japanese economic performance. This suggests that comparative advantage based on natural resource endowments and traditional skills can relatively quickly

67. Bronfenbrenner (1971), pp. 58-9.

be lost to countries more willing to develop on and diversitfy from existing bases of knowledge. Once comparative advantage is lost, both growth and the balance of payments will suffer. Comparative advantage in innovation need by no means, however, be found only by looking at existing patterns of activity. The accidents of birth and history which brought a Bessemer or Marconi into Germany and the USA gave these nations a head start in the development of furnaces and radio equipment. Governments cannot influence the random occurence of inventive talent, but they can do their best to stimulate and support it when it arrives. (There are, of course, problems here for world welfare and, indirectly, for the country itself. Much innovative talent works to maximum effect when teamed with complementary inputs — and if these are only available abroad, as is often the case for smaller countries, there may be no way of realising on potential of talent if it is kept at home).

Looking at the question from a different angle, it is clear that the growth objective may well suffer if governments become excessively preoccupied with balance of payments problems. In this connection, Pollard (1982) has argued that Britain's poor growth performance in the last two decades is the result of "stop-go" policies in the 1950s and 1960s. During the "stop" phases of the cycle, he argues, businessmen were discouraged from investing in new equipment by policies aimed at eradicating payments deficits which had arisen in the course of a preceding "go" phase. What is of relevance here is that a declining willingness to invest is associated with (a) a slower rate of acquisition of equipment embodying new technology, and (b) a slower rate of investment in technological knowledge. The long-run foundations for future growth are thus undermined.

By way of conclusion, it is perhaps as well to discuss whether policies supposed to operate through the rate and direction of technical progress are likely to be substantially different in their characteristics from the standard instruments of demand management. The latter are generally aimed at a short-run horizon, are regarded as being reasonably predictable in effect, and, by definition, seek to exert influence on aggregate demand side variables. Technology policy can only be aimed at a more distant horizon, is for that reason and others much less predictable in its effects, and would seem to be principally concerned with the supply side.

Such a dichotomy is probably more apparent than real, however. All short-term, cycle-oriented, demand management policy aimed at stimulating (or suppressing) economic activity has the potential for speeding up (or slowing down) both innovation and the diffusion of new products and techniques. An example of this was discussed in relation to the balance of payments objective. At the same time, the unceasing growth of productivity associated with successful innovation may require demand side stimulation in relation to expenditure on both new and existing goods if a given quantum of employment is to be maintained period by period. And even without necessarily subscribing to the notion of regular Schumpeterian innovation cycles, we have no reason to expect that, over time, innovations will appear at a constant or constantly growing rate, each one with similar implications for changes (or potential changes) in consumption and investment demand. Any irregularities may well bring with them disturbances to growth trends in aggragete demand. After decades of neglect, the intimate and integral dynamic relationship between the demand and supply sides in the diffusion of innovations has only recently been receiving the attention it derserves[68]. At its simplest, this relationship draws attention to the way in which expanding demand permits static and dynamic scale economies to be reaped on the supply side — with a resultant fall in prices and further expansion of demand in consequence. More complex versions of this vision indicate how the number and variety of goods and services supplied are responses to experiment in and feedback from the market. Almost any demand side policy can therefore have implications for the rates of innovation and diffusion, and any policy aimed at the innovation rate directly may, if it has any impact at all, reverberate eventually on aggregate demand.

REFERENCES

ABERNATHY, W.J. and TOWNSEND, P.L. (1975), "Technology, Productivity and Process Change", *Technological Forecasting and Social Change.*

ARROW, K.J. (1962), "Economic Welfare and the Allocation of Resources for Invention", *The Rate and Direction of Inventive Activity : Economic and Social Factors*, N.B.E.R., pp. 609-626.

—— (1963), *Social Choice and Individual Values*, Wiley.

—— and HAHN, F. (1971), *General Competitive Analysis*, Oliver and Boyd.

BARZEL, Y. (1968), "Optimal Timing of Innovations", *Review of Economics and Statistics,* vol. 50.

BLISS, C.J. (1975), *Capital Theory and the Distribution of Income*, North Holland.

BRONFENBRENNER, M. (1971), *Income Distribution Theory*, MacMillan.

68. See, for example, Pasinetti (op. cit.), Mowery & Rosenberg (op. cit.), Metcalfe (1981) and Stoneman & Ireland (1983).

DASGUPTA, P. and STIGLITZ, J. (1980), "Uncertainty, Industrial Structure and the Speed of R & D", *Bell Journal of Economics*, vol. 11.

DEMSETZ, H. (1969), "Information and Efficiency: Another Viewpoint", *Journal of Law and Economics*, vol. 12, reprinted in Lamberton, D.M. (ed.) (1971) *Economics of Information and Knowledge*, Penguin.

ENOS, J. (1958), "A Measure of the Rate of Technological Progress in the Petroleum Refining Industry", *Journal of Industrial Economics*, vol. 6.

FIRESTONE, O.J. (1971), *Economic Implications of Patents*, University of Ottawa Press.

FREEMAN, C. (1980), "Government Policy", in Pavitt, K. (ed.), *Technical Innovation and British Economic Performance*, MacMillan.

GANNICOTT, K. (1980), "Research and Development Incentives" in the Myers Committee Report, *Technological Change in Australia*, (vol. IV), Australian Government Publishing Service.

GRAAF, J. de V. (1975), *Theoretical Welfare Economics*, Cambridge University Press.

GREATER LONDON ENTERPRISE BOARD (1983), *Technology Networks*.

HALL, P.H. and HEFFERNAN, S. (1985), "More on the Employment Effects of Innovation", *Journal of Development Economics* vol. 17 (1-2) (forthcoming).

JOHNSTON, R. and GUMMETT, P. (1979), *Directing Technology : Policies for. Promotion and Control*, Croom Helm.

KAMIEN, M.I. and SCHWARTZ, N.L. (1982), *Market Structure and Innovation*, Cambridge University Press.

KEMP, M. (1980), Review of Dasgupta, P.S. and Heal, G.M. *Economic Theory and Exhaustible Resources*, *Economic Journal*, vol. 90, p. 938.

LAMBERTON, D.M. (ed.) (1971), *Economics of Information and Knowlegde*, Penguin.

LIPSEY, R.G. and LANCASTER, K. (1956), "The General Theory of Second Best", *Review of Economic Studies*, vol. 24.

LOWE, J., *Innovation and Technology Transfer* (forthcoming).

MACDONALD, S. and LAMBERTON, D. (1981), *Tradition in Transition : Technological Change and Employment in Banking,* Working Paper 33, Department of Economics, University of Queensland.

MANDEVILLE, T., LAMBERTON, D. and BISHOP, E. (1982), *Economic Effects of the Australian Patent System*, Australian Government Publishing Service.

MANSFIELD, E. *et al.* (1971), *Research and Innovation in the Modern Corporation*, MacMillan.

——, SCHWARTZ, M. and WAGNER, S. (1981), "Imitation Costs and Patents", *Economic Journal,* vol. 91, December.

MEADOWS, D. (1968), "Estimate Accuracy and Project Selection Models in Industrial Research", *Industrial Management Review*, vol. 9.

METCALFE, J.S. (1981), "Impulse and Diffusion in the Study of Technical Change", *Futures* (October).

——, and HALL, P.H. (1983), "The Verdoorn Law and the Salter Mechanism: A Note on Australian Manufacturing Industry", *Australian Economic Papers*, vol. 22.

MOWERY, D.C. and ROSENBERG, N. (1979), "The Influence of Market Demand upon Innovation: A Critical Review of Some Recent Empirical Studies", *Research Policy*, vol. 8, April.

MYERS, S. and MARQUIS, D. (1973), *Successful Industrial Innovations*, National Science Foundation, Washington.

NEARY, J.P. (1981), "On the Short-run Effects of Technological Progress", *Oxford Economic Papers*, vol. 33.

NELSON, R.R. and WINTER, S.G. (1982), *An Evolutionary Theory of Economic Change*, Belknap, Harvard.

NG, Y-K. (1979), *Welfare Economics*, MacMillan.

NORDHAUS, W.D. (1969), *Invention, Growth and Welfare: A Theoretical Treatment of Technological Change*, M.I.T.

PASINETTI, L. (1981), *Structural Change and Economic Growth*, Cambridge University Press.

PAVITT, K. and WALKER, W. (1976), "Government Policies Towards Industrial Innovation: A Review", *Research Policy*, vol. 5.

PHELPS, E.S. (1965), *Fiscal Neutrality towards Economic Growth*, McGraw-Hill reprinted, in part, as Chapter 21 of Sen, A. (ed.) (1970) *Growth Economics*, Penguin.

——, (1966), "Models of Technical Progress and the Golden Rule of Research", *Review of Economic Studies*, vol. 33, April.

POLLARD, S. (1982), *The Wasting of the British Economy: British Economic Policy, 1945 to the Present*. Croom Helm.

RONAYNE, J. (1984), *Science in Government*, Edward Arnold.

ROSENBERG, N. (1976), *Perspectives on Technology*, Cambridge University Press.

—— (1982), *Inside the Black Box: Technology and Economics*, Cambridge University Press.

SALTER, W.E.G. (1966), *Productivity and Technical Change*, Cambridge University Press.

SAMUELSON, P.A. (1954), "The Pure Theory of Public Expenditure", *Review of Economics and Statistics*, vol. 36, November.

SCHMOKLER, J. (1966), *Invention and Economic Growth*, Harvard.

SCHUMPETER, J.A. (1934), *The Theory of Economic Development*, Harvard.

SCIENCE POLICY RESEARCH UNIT (1972), *Success and Failure in Industrial Innovation* (Project SAPPHO), Centre for the Study of Industrial Innovation, London.

SIMON, H.A. (1955), "A Behavioural Model of Rational Choice", *Quarterly Journal of Economics*, vol. 69.

—— (1959), "Theories of Decision Making in Economics", *American Economic Review*, vol. 49.

SINCLAIR, P.J.N. (1981), "When Will Technical Progress Destroy Jobs?", *Oxford Economic Papers*, vol. 33.

STONEMAN, P. and IRELAND, N.J. (1983), "The Role of Supply Factors in the Diffusion of New Process Technology", *Economic Journal* (Supplement).

TAYLOR, C.T. and SILBERSTON, Z.A. (1973), *The Economic Impact of the Patent System*, Cambridge University Press.

THOMAS ,H. (1971), "Some Evidence on the Accuracy of Forecasts in R & D Projects", *R & D Management*, vol. 1.

TISDELL, C. (1981), *Science and Technology Policy : Priorities of Governments*, Chapman and Hall.

USHER, D. (1980), *The Measurement of Economic Growth*, Blackwell.

UZAWA, H. (1965), "Optimum Technical Change in an Aggregate Model of Economic Growth", *International Economic Review*, vol. 6, January.

WATKINS, D.S. and RUBENSTEIN, A.H. (1979), "Decision-Makers' Responses to Industrial Innovation Incentives" in Baker, M.J. (ed.), *Industrial Innovation*, MacMillan.

WRIGHT, B.D. (1983), "The Economics of Invention Incentives: Patents, Prizes and Research Contracts", *American Economic Review*, vol. 73, September.

CHAPTER TWO

TECHNOLOGICAL INNOVATION AND THE COMPETITIVE PROCESS

By J. S. Metcalfe*

It has long been recognized that the economic impact of new technology, and thus its consequences for standards of living, the structure of economic activity and patterns of foreign trade, is dependent upon the rate and direction with which that technology is absorbed into the prevailing economic environment. It is also recognized that this process of absorption is dependent on the economic advantages which the new technology offers relative to the existing technologies which are available. It is perhaps less well recognized that this absorption process is itself a key determinant of the direction in which technologies evolve. The purpose of this paper is to explore the interrelationship between economic absorption, differential economic advantage and induced technical change in terms of a treatment of the competitive process which is built around three logically separate, but in practice tightly interwoven mechanisms; namely, the diffusion mechanism, the selection mechanism and the inducement mechanism, When drawn together the three strands of analysis suggest that the diffusion of a technology and the development of that technology are simultaneously determined within any competitive market economy.

A. COMPETITION AND TECHNOLOGY

Before proceeding further it will prove useful to clarify the meaning

* University of Manchester and University of New England
This paper is a formal elaboration of ideas derived within an ESRC/SERC Joint Committee funded project on Post Innovative Performance in U.K. Industry. I am much indebted to my colleagues Professor Mike Gibbons, Luke Georghiou, Janet Evans and Tim Ray for stimulus and helpful discussion. Needless to add, I alone am responsible for any errors evolved in the translation of ideas into the language of economics. A book based on our empirical research is to be published by Macmillan in 1985 under the provisional title "Post Innovative Performance". I am also much indebted to the University of New England for the award of a Visiting Research Fellowship in the Department of Economics, which enabled me to draft this paper and enjoy the benefit of close discussion with the editor of this symposium.

of the terms "competition" and "technology" as employed in the ensuing argument.

Within economic theory, the notion of competition covers a spectrum of ideas. Most familiar among these is "atomistic competition" which is itself dependent on two separate concepts. On the one hand, the firm is perceived as a price taker in input and output markets, while, on the other hand, there is freedom of entry into the market. It is price taking which corresponds to there being many firms in the market and freedom of entry which ensures that firms earn a normal rate of return on the capital invested in production activity. Invaluable to the development of theories of resource allocation and market structure, the notion of atomistic competition is, however, severely circumscribed by its focus upon states of market and firm equilibrium, as distinct from the process by which those states are obtained. As Hayek has pointed out, atomistic competition cannot be meaningfully related to the verb "to compete". Moreover, it provides little or no guidance on the relation between the forces of competition and the discovery of the new resources, new technologies and new institutions, which are an integral part of economic growth and development[1].

By contrast in this paper we emphasize the notion of "innovative competition", as an active process of rivalry between firms. In common with atomistic competition, it subsumes the elements of open competition, interpreted as the absence of barriers to entry, actual or potential, to a given production activity[2]. But more than this, innovative competition is based upon the notion of differentiation as the chief means by which firms gain market advantages relative to their rivals. Not entry, *per se*, but entry on advantageous terms related primarily to success in developing new technology is the *sine qua non* of innovative competition. None other than Schumpeter's perennial gale of creative destruction, it is the quest for differential advantage which drives the competitive process and which determines the conditions of interfirm rivalry for market share and command over production resources[3]. Not entry and co-existence but entry and market dominance is the theme to which competition is played in modern industrial society, and technological

1. P.J. McNulty, 1968, "Economic Theory and the Meaning of Competition", *Quarterly Journal of Economics*, vol. 82, pp. 639-656. The Hayek reference is to "The Meaning of Competition", in F.A. Hayek, 1964, *Individualism and Economic Order*, Chicago.

2. P.W.S. Andrews, 1964, *On Competition in Economic Theory*, Macmillan, p. 16.

3. J. Schumpeter, 1947, *Capitalism, Socialism and Democracy*, George Allen and Unwin. See also, B.R. Williams, 1949, "Types of Competition and the Theory of Employment", *Oxford Economic Papers*, vol. 1, where it is argued that "the special mark of the fully dynamic problem is the difference between different individuals".

innovation is the principle plot. Such innovative competition is inherently a process and not a state. Yet as Schumpeter and Hayek both emphasized, it may well result in the long run in states which are competitive in terms of the profits which are earned, while being non-atomistic in terms of the structure of the industry.

In this perspective, the study of competition as a process can proceed at two distinct levels : analysis of the consequences of given differences between rival firms; and analysis of the procedures and strategies by which firms gain a differential advantage. Of course, firms are inevitably complex institutions and it is fortunate that not all the potential sources of inter-firm differentiation are relevant to our enquiry. Our concern is primarily with those differences which give rise to different levels of profitability and which enable firms to accumulate capacity and market share to the disadvantage of their rivals. We shall visualize each firm as an island of privileged information, with a knowledge base and an organizational capacity largely determined by its past history. The important differences between firms can then be located in three distinct categories.

First, differences in the process of production and the quality of output, which relate directly to the firm's knowledge base, the individuals involved and the organizational structure by which their efforts are coordinated. It is these elements which primarily determine a firm's profit margin, the basis for improvements in its competitive position.

Second, we have those differences which relate to the translation of profits into the accumulation of production capacity and the growth of market share at the expense of rival firms. Crucially important here are the capacities of firms to command finance in excess of internally generated profits, and the ability to grow without impairing the efficiency with which current production activity is undertaken[4]. Here we find much of the substance of business history : the firm whose owners and managers do not wish it to grow beyond a certain size; the firm with technological advantages but unable to command the external finance to turn superiority into competitive advantage; the firm whose investment strategies are constrained by its membership of a larger corporate group; and the firm which courts bankruptcy by its over extended attempts to grow.

Third, we have the most significant differences of all, namely those

4. Cf., E.T. Penrose, 1961, *The Theory of the Growth of the Firm,* Blackwell. Also N. Kaldor, 1934, "The Equilibrium of the Firm", *Economic Journal,* vol. 44, pp. 60-76, who first advanced the thesis that entrepreneurial/managerial costs are related to the rate of expansion of the firm and not its scale.

relating to the ability to innovate, in technological or organizational terms, and create new differential advantages. So many important factors are involved here that it is impossible to be comprehensive, but any worthwhile listing must include : differences in the ability to "read" the scientific and technological environment; differences in the resources informally and formally devoted to research and development activity; differences in the efficiency with which research resources are managed; and differences in the effectiveness with which research activities are related to the other marketing and production activities of the firm. Such a list is immediately recognizable within the literature on success and failure in the innovation process[5], but it will readily be admitted that differences between firms of this nature are extraordinarily difficult to incorporate within the corpus of static competitive theory. What is needed, is a framework in which differences between firms in the technologies they command is reflected in their relative position and growth in the markets for the products they produce. To provide the basis for such a framework is our purpose in this paper.

At this point it is convenient to turn from questions of competition to questions of technology and the problems of breaking into the black box which typifies economists' endeavours in the field[6]. At a general level we can define technology in terms of knowledge, the knowledge to design goods and services to attain certain objectives. At a specific level we then identify any given technology by that set of design principles, based on technological or scientific understanding, which identify particular commodities and their processes of production. For any particular commodity, these design principles are reflected in, and may be quantified in terms of, a set of performance characteristics. In turn these may be divided into process characteristics — specified quantities of materials and energy together with the requisite physical transformations required to produce a unit of the commodity — and product characteristics — the functions which the commodity fulfils[7]. When different commodities embody a common set of design princi-

5. C.F. Carter and B.R. Williams, 1959, *Industry and Technical Progress,* Oxford, J. Langrish (et al.) 1969, *Wealth from Knowledge,* Macmillan, and R. Rothwell and W. Zegveld 1982, *Innovation and the Small and Medium Sized Firm,* Pinter.

6. Cf., N. Rosenberg, 1982, *Inside the Black Box,* Cambridge.

7. Process characteristics are measured in principle in any form of input : output analysis. The converse role of product characteristics has been made clear in the Ironmonger/Lancaster theory of consumer behaviour. D. Ironmonger, 1972, *New Commodities and Consumer Behaviour,* Cambridge, and K. Lancaster, 1971, *Consumer Demand : A New Approach,* Colombia. Of course, the characteristics approach applies to producer goods just as readily as to consumer goods.

ples we shall say they belong to the same design configuration. Within a design configuration each commodity is located in a space of performance characteristics. It is one of the chief attributes of any configuration that the attained set of performance characteristics evolves over time. A configuration is rarely, if ever, fully developed at the date the particular design principles are first established in an innovation. Rather one observes a sequence of innovations exploring the agenda of the configuration, "the children of the original innovation"[8], gradually improving the levels of performance attained by the various characteristics. For each innovation and for each design configuration it is normally possible to identify, in advance, limits to performance characteristics: a position of technological maturity implicit in the particular set of design principles, (the chemical and physical properties of materials and the laws of thermodynamics) and their interrelation. These design limits have an important bearing upon the economics of maturing technologies as we shall see further below.

This scheme may be taken one stage further by recognizing situations in which different design configurations satisfy the same set of user needs. We then have what we shall call a technological regime. It is characteristic of a regime that it draws upon different bodies of design principles for its development, as may be illustrated by example. The requirement to transport people and cargo over the oceans has been met historically by several design configurations, which may be loosely grouped into the sailing ship, the steamship and the motor ship, although in practice there are many distinct configurations within each broad group. The requirement to communicate effectively over distance, ie. independently of face-to-face contact, is associated with the telegraph, radio, telephone and satellite communications. And, as a final example, the requirement to clothe people against the rigours of climate has lead to design configuration based upon natural and synthetic fibres. In each of these cases the regimes have developed through a lengthy historical process which is manifested in a sequence of innovations, each one augmenting the performance of its particular configuration.

The purpose of these distinctions is of course, to put them to practical use in understanding the relationships between technical change and the competitive process. Within their confines, for example, we can identify two distinct types of technical progress: improvements that occur within a design

8. Professor J. Hicks', felicitious phrase, see his, 1977, *Economic Perspectives*, Oxford, pp. 16-17. Improvement in some dimensions of characteristics space may of course, imply deterioration in other dimensions, so that only a "global" evaluation of the technology can indicate the rate and direction of progress.

configuration; and the addition of new design configuration within a technological regime. The sources and thus the costs of technological knowledge will differ between these cases, and it is to the latter that we must look if technical progress is not to slacken with the passage of time. More importantly for present purposes, these distinctions enable us to identify three distinct levels of technological competition, within a configuration, between configurations but within a regime, and between regimes. Between these levels the market environment will differ as will the knowledge base of the different rivals. In this paper we will be concerned solely with a regime which consists of a single design configuration, and we shall investigate two dimensions of the competitive process, namely the absorption of the configuration into the existing economic structure, and the process of competition within the configuration as a dominant design emerges from the rival innovations.

To treat the complex of factors which are necessarily involved in the process of innovation competition is not easy, and we shall find it useful to divide our argument into the following stages. We begin with a design configuration that consists of a single innovation with given, time invariant, performance characteristics, and we investigate the process by which it is diffused into an equilibrium economic niche. Since we ignore all questions of foreign trade, this process necessarily involves a balancing of the growth of the market with the growth of the capacity to produce the innovation. We next allow many innovations to form the design configuration and investigate the process of competition in terms of rivalry between the different but given innovations, each one "championed" by one or more firms.

Finally we turn to the inducement mechanisms which improve the performance of each innovation under the pressures and constraints of the diffusion and selection mechanisms. This final section leads us to the conclusion that the diffusion and development of a technology are inseparable phenomena.

B. THE DIFFUSION MECHANISM

We begin with an outline of a diffusion process for a single innovation based upon the simultaneous development of market demand and the accumulation of productive capacity for a non-durable commodity. Our central idea is that the development of the new technology acts as an impulse to growth and structural change and that this impulse can usefully be analysed via a distinction between the long-period niche for the new technology and the process of adjustment into that niche.

The long period niche will depend on the given performance characteristics of the new commodity relative to existing complements and substitutes, on its relative price, and upon a miscellaneous list of elements, such as per capita income, which influence demand. Some of these determinants may change exogenously over time, so altering the niche but we ignore such influences at this stage. More significant will be changes in the price of the new commodity, p, which will be determined endogenously by the diffusion process. Let n be the equilibrium demand for the commodity measured in physical output units. We postulate that $n = n(p)$ and more specifically and for analytic convenience alone, $n = c - ap$, with c, a, constants determining this long-period demand curve.

It is central to a Schumpeterian development process that the long-period position is not attained immediately the initial innovation is made, and one reason of this delay rests in the uncertainty surrounding the characteristics of the new commodity. Potential consumers have to acquire information and learn new preference patterns before they purchase the new commodity[9]. As Hoffman put it, an industry may experience a rapid rate of growth not because of fortuitous Engel Law effects but,

> "because it is a 'young' industry busy creating a
> market for itself in the place of other products;
> in contrast to an 'old' industry where the public
> has had a long time to decide on how much to use"[10].

Many diffusion studies emphasize a process of social learning, in which individuals gain knowledge by observing the experience of existing consumers. Demand then grows at a rate dictated by such factors as the nature of communication channels, the distribution of user receptivity to new commodities, and the degree of technical sophistication of the new commodity. Typically the learning process generates a sigmoid growth curve of demand and with no loss of generality we assume that the logistic curve is an adequate summary of the demand generating process[11].

9. Cf. L. Pasinetti, 1981, *Structural Change and Economic Growth*, Cambridge, pp. 75-76.

10. W.G. Hoffman, 1949, "The Growth of Industrial Production in Great Britain", *Economic History Review*, vol. 4.

11. For up-to-date surveys of the diffusion literature consult S. Davies, 1979, *The Diffusion of Process Innovations*, Cambridge, and P. Stoneman, 1983, *The Economic Analysis of Technological Change*, Oxford. In this context one may recall the comment attributed to E. Wellbourne (Master of Emmanual College, Cambridge), "When a new thing is invented, some men will pay more for it than it is worth, because it is new; other men

If $x(t)$ is the rate of demand at time t, then

$$\frac{dx}{dt} = \beta x(t) \left[n(p) - x(t) \right] \qquad (1)$$

where β is the adoption coefficient, reflecting the information transmitting mechanism, and $n(p)$ is the equilibrium demand curve. It will be clear from (1) that we have not a single diffusion curve but a complete family of diffusion curves each one defined relative to a given price for the new commodity. To understand the diffusion process we must determine this price and the forces which cause it to change and generate *a diffusion envelope* in place of a single diffusion curve. This brings us directly to the second dynamic element, the accumulation of capacity to produce the new commodity.

We begin with the proposition that the growth of capacity depends on the rate of profits. For present purposes we adopt the simple postulate that capacity expansion continues as long as the expected rate of profit exceeds an appropriate long-period 'normal' level, $\bar{r}$, and that the expected rate of profit is equated with the current rate of profit, r. Funds to finance capacity expansion come from two sources, internal funds corresponding to the net profits of firms in the industry, and external funds, in which category we include the entry of new firms. As a first approximation we summarize financial conditions in terms of a fixed fraction, f, of industry profits reinvested in capacity expansion, and a fixed ratio, e, between internal and external funds. These ratios are the same for all firms. We then have a version of the classical savings postulate, with the growth rate of capacity, g, related to the rate of profits by

$$g = f(l + e)r = \Pi r \qquad (2)$$

All firms in the industry, irrespective of their date of entry, operate a common process technology with equal efficiency to produce the identical product. This process is a constant returns to scale process with capital output ratio, v, and unit primary input requirement, l, where, again for expositional purposes, we assume that labour time is the only non-capital input. If w is the wage rate per unit of labour time then the net rate of profits is given by

$$r = \frac{p - wl}{v} - (\bar{r} + d)$$

will not buy it at all unless you give it to them — and for essentially the same reason". Quoted from C. Northcote Parkinson 1966, *A Law Unto Themselves*, John Murray, 1966, p. 39.

with d the given exponential rate of depreciation. To complete the supply side we postulate that the reproducible capital goods are obtainable at a given price and that the wage rate is an increasing function of the volume of employment and thus of the scale of output, hence $w(x) = w_0 + w_1 x$. Combining these elements together yields the fundamental relation between the growth rate of output and the price of the new commodity

$$p = kg + h_0 + h_1 x \qquad (3)^{12}$$

At a given price it follows from (3) that the growth of output will also follow a logistic curve toward a long-run niche, at which the rate of profits is driven to the long-period normal level. However, it is clear, on comparing (3) with (1), that the respective long-period niches for demand and capacity will not be the same and that the path of demand diffusion will differ from the path of capacity diffusion. Clearly an arbitrarily determined price is untenable for it involves situations of excess or deficient capacity during the growth process. Profit conscious entrepreneurs will not tolerate such lost opportunities and we therefore postulate that the price of the new commodity adjusts to maintain a continuous equality between the growth of demand and the growth of capacity. Such a diffusion path we call a balanced path, and we treat it as the secular trend of industrial growth. In the practical course of growth, it is inevitable that mistakes will be made and that actual prices and outputs will deviate from the path of balanced expansion. However, our working hypothesis is that any deviations are sufficiently minor and short-lived as to make the balanced path representative of the secular course of development.

On equating (3) with (1), we find that the growth rate and the balanced price are related to the scale of output by the relations

$$g(t) = \frac{\beta(1 + ah_1)}{1 + a\beta k} \left[\frac{c - ah_0}{1 + ah_1} - x(t) \right] \qquad (4)$$

and

$$p(t) = \frac{k\beta c + h_0}{1 + a\beta k} + \left[\frac{h_1 - k\beta}{1 + a\beta k} \right] x(t) \qquad (5)$$

The growth rate decline as output expands, and price rises or falls with x, according as the effect of a rising wage rate on unit costs exceeds or falls

12. $k = v/\Pi$, $h_0 = w_0 l - (\overline{r} + d)v$, and $h_1 = w_1 l$.

short of the effect of the declining growth rate on profit margins. On integrating (4) we find that the secular path of output is a logistic curve

$$x(t) = \frac{C}{1 + Ae^{-BCt}}$$

with $\quad C = \dfrac{c - ah_0}{1 + ah_1}$, $B = \dfrac{\beta(1 + ah_1)}{1 + a\beta k}$ and $A = \dfrac{C - z}{z}$ (6)

where z, is the initial level of output $x(0)$[13].

In the above, C, is the equilibrium long-period niche for the new commodity, and B, is the diffusion coefficient, both of which depend on the parameters governing the growth of demand and capacity[14]. Figure 1 illustrates the main aspects of the process whereby the innovation is absorbed into the prevailing economic structure. Figure 1a shows the secular pattern of expansion toward the long-period niche. It is an envelope curve at each point on which the price and cost conditions in the industry are different. Figure 1b shows the time path of the output growth rate and we see that this is subject to continuous retardation. In Figure 1c we show the paths of price and unit costs (with $h_1 > k\beta$), indicating how unit profit margins are squeezed during the diffusion process. The rate of profits on invested capital clearly declines along the balanced path and we are reminded of the transient nature of innovation based profit and Schumpeter's incisive statement that "it is at the same time the child and the victim of development"[15]. The annual flow of profit accruing to the entire industry is shown in Figure 1d, it is the product of the rate of profits on capital, which continually declines toward $\bar{r}$, and a capital stock which expands towards a long-period level. The combination of effects gives the bell-shaped curve. The time profile of investment will exactly match the time profile of this profit curve.

It is characteristic of this framework that the diffusion environment changes continuously and endogenously as a direct result of the diffusion process. Expansion increases unit costs, and the growth rate determines the required profit margin so the price changes continuously from within.

13. Non-linear demand and input supply schedules will produce a non-logistic but still sigmoid balanced path.

14. The relationship of (6) with the standard diffusion curve is discussed in J.S. Metcalfe, 1981, "Impulse and Diffusion in the Study of Technical Change", Futures, vol. 13. Reprinted in C. Freeman, 1984, *Long Wawes in the World Economy*, Butterworth.

15. J. Schumpeter, 1934, *The Theory of Economic Development*, Oxford, p. 154.

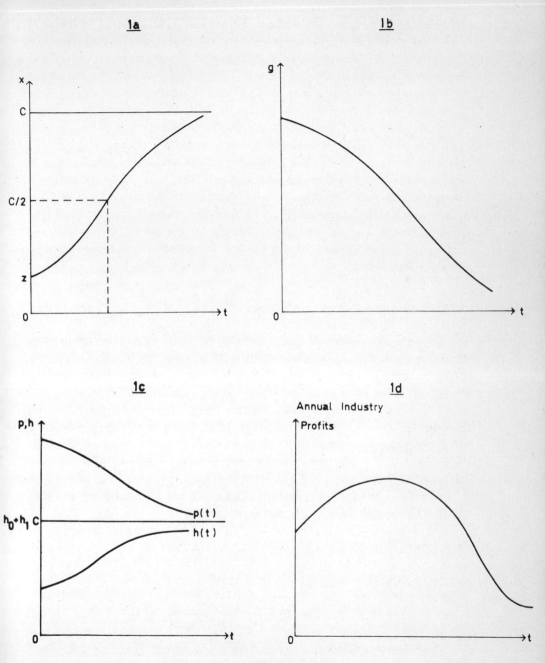

FIGURE 1

As Bela Gold has forcefully remarked, the equilibrium level of demand at the initial stages of the diffusion process may then greatly understate the equilibrium level which defines the final niche[16]. This framework can readily incorporate the consequences of exogenous changes in the factors determining diffusion. The impact of a change in the demand curve, or in input supply, or in the financial conditions, for example, can always be treated in terms of the impact on the parameters B and C and thus the effects on output price and profit paths. A sequence of such exogenous changes may then be run together to determine the path of diffusion within an environment of continuous change. It will not be a logistic path but that is of no consequence. For present purposes, however, these are not questions of great significance. More important is the need to abandon the assumptions that all firms employ the same technology and that this technology is unchanged throughout the diffusion process. To these questions of diversity within the design configuration we now turn.

C. THE SELECTION MECHANISM

We now allow there to be many different, rival innovations within the design configuration, the total number being given with each innovation distinguishable in terms of its given product and process characteristics. The questions we wish to answer relate to the economic mechanisms which select between rival innovations and the effect of technological diversity upon the diffusion path. We begin with the case in which firms produce the same homogeneous commodity using different process technologies, and then extend this to incorporate differences in product technology. We continue to identify the innovating unit with the firm, *simpliciter*. It is of no consequence for our present purposes that we assume that firms using the same process technology do so with equal efficiency.

(i) Differences in Process Technology

Each process innovation is of the same constant returns kind, with homogeneous labour and a machine the only inputs. For expositional purposes we shall treat all input prices as independent of the scale of the industry and exogenously given for the purposes of analysis. Each firm is identified with a particular process technology and builds its accumulation

16. B. Gold, 1981, "Technological Diffusion in Industry : Research Needs and Shortcomings", *Journal of Industrial Economics*, vol. 24.

plans around investment in that process. At the given input prices the different processes operate with different levels of unit cost, differences which
are the basis for a competitive selection process. Diversity in unit costs,
when translated into diversity in profit margins provides the basis for the
more efficient to expand market share at the expense of the less efficient.
Since our concern is with the consequences of technological diversity as a
basis for competition, we shall assume that all firms have the same propensity to accumulate, k, defined as the ratio between unit profit margin and the
growth rate of the firms' capacity output. The consequences of inter-firm
differences in the propensity to accumulate are treated briefly in a caveat
to the main argument.

If h_i is the level of unit cost in firm i, its capacity growth rate is given by

$$g_i = \Pi_i r_i = \frac{[p - h_i]}{k} \tag{7}$$

with $k = v_i / \Pi_i$, the same for all firms within the design configuration. Equation (7) is the appropriate generalization of (3) to each individual firm and
its particular process technology. From this we see that the growth rate of the
productive capacity invested in each technology is proportional to the corresponding profit margin, with the profit rates and growth rates differing between the firms and their technologies.

The growth rate for the industry as a whole, is the weighted sum of the
capacity growth rates of the individual technologies. Let s_i be the share of
technology i in total output, then $g = \Sigma s_i g_i$ or

$$g = \frac{1}{k} [p - \bar{h}] \tag{8}$$

where $\bar{h} = \Sigma s_i h_i$ the average practice level of unit cost[17]. It is clear from (8)
that the industry growth rate is proportional to the average rate of return on
capital invested in the industry.

The immediate point of interest here is the link between growth rate,
price and average practice unit costs. The literature on technical change
has long realized the empirical importance of differences between average
practice unit costs (or what amounts to the same thing, average practice pro-

17. For reasons which we make clear in footnote 20, the summation in (8) is taken
only with respect to firms and technologies which have a positive profit margin and rate
of growth.

ductivity) and best practice unit costs. Here we have a framework in which this distinction has analytic significances in the context of the competitive process[18].

To explore this competitive process in greater detail it is helpful to identify the possible states in which a firm and its technology may be located at any given point in time. These states, which depend upon the relative configurations of p, $\bar{h}$ and h_i, may be enumerated as follows :

a) $p > \bar{h} > h_i$. This firm utilizes a technology which, at the given input prices is of above average efficiency. Such a firm grows faster than the average for the industry as a whole, and consequently increases the economic weight of its technology as measured by market share.

b) $p > \bar{h} = h_i$. A firm which is dynamically representative of the industry as a whole, growing at the industry average rate with a constant market share for its technology.

(c) $p > h_i > \bar{h}$. A firm with above normal profits although it is of below average efficiency. This firm will be growing in absolute scale but its growth rate is below the industry average so it is losing market share.

(d) $p = h_i > \bar{h}$. The marginal firm which is no longer able to grow, earning only the normal rate of profits r. Such a firm must continually lose market share and may also be in absolute decline.

(e) $p < h_i$. The bankrupt or non-viable firm. Although, strictly speaking, a firm will continue to produce as long as prime costs are covered (h_i, includes capital changes it will be remembered) we shall employ the device of assigning zero output to all bankrupt firms.

Figure 2 may help the reader to translate the possible states into more familiar terms. The diagram relates to a particular moment in time and along the axes are measured the unit input requirements of machines, a_1 and labour a_2. A point identifies the production method of an individual firm and more than one firm can occupy the same point. The density of firms within this space is arbitrary and depends on the initial conditions. All firms are contained within the boundaries of the convex hull labelled $a - a$.

At a given point in time the average practice technical coefficients are $\bar{a}_1$ and $\bar{a}_2$, they depend on the market shares of the producing firms and their associated input : output coefficients. Given relative input prices, we identify

18. Cf. W.E.G. Salter, 1961, *Productivity and Technical Change*, Cambridge, and R.R. Nelson, 1981, "Research on Productivity Growth and Productivity Differences: Dead Ends or New Departures", *Journal of Economic Literature*, vol. 19.

average practice unit cost by the line $\overline{h} - \overline{h}$ and best practice unit cost by the line $h_\alpha - h_\alpha$, a being the best practice process at these prices. This gap between average and best practice will change over time under the combined forces of the diffusion and selection processes.

As the industry is assumed to be expanding, price exceeds $\overline{h}$ and this margin determines which firms are viable at that particular date. Firms in region A are no longer viable, firms in B are growing but losing market share, while firms in C are expanding both absolutely and relatively. The line $p - p$ stands above $\overline{h} - \overline{h}$ by the distance kg, and firms distributed along it are marginal. Firms distributed along $\overline{h} - \overline{h}$ are dynamically representative.

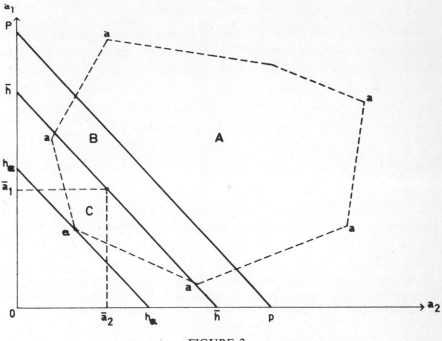

FIGURE 2

Consider now the operation of the selection mechanism, assuming that the growth rate of demand, g_D is given and that the industry is on a path of secular balance, as in the previous section. Then $kg_D = p - \overline{h}$ but, since the various firms have different growth rates, $\overline{h}$ must be falling over time as the more efficient technologies increase their relative market share and acquire a greater economic weight. It follows that the price along the balanced path falls at the same rate as average practice unit cost even though g_D is here constant. This process continually enlarges section A of Figure 2, and takes an increasing number of technologies into marginality and beyond into non-viability[19]. The mechanism by which $\bar{h}$ falls is as follows.

From the definition of $\bar{h}$ it follows that $d\bar{h}/dt = \Sigma(ds_i/dt)h_i$. For every firm $ds_i/dt \equiv s_i(g_i - g)$ and so from (7) and (8) we find

$$\frac{ds_i}{dt} = \frac{s_i}{k}[\bar{h} - h_i], \ h_i < p \tag{9}$$

Hence,

$$\frac{d\bar{h}}{dt} = \frac{1}{k}\sum_i s_i(\bar{h} - h_i)h_i = -\frac{1}{k}V(h, t) \tag{10}$$

where $V(h, t)$ is the variance of unit costs at time t. The summation in (10) only applies to those firms satisfying $p_i > h_i$, since selection only operates with respect to the viable firms. By a similar argument we find that

$$\frac{d^2\bar{h}}{dt^2} = \frac{-1}{k^2}S(h, t) \tag{11}$$

where $S(h, t)$ is the third moment of the distribution of h_i around average practice. The sign of $d^2\bar{h}/dt^2$ may change several times during the selection process depending on how the distribution of shares evolves with time. Equations (9), (10) and (11) are the fundamental equations of the selection process, and indicate that the rate of decline of average practice unit cost is proportio-

---

19. At this point we must refer to the problem alluded to in footnote 17, namely the exclusion of marginal firms from the definition of $\bar{h}$ and g. The growth rate of total output is not determined solely by the growth rate of profitable firms but by this in conjunction with the growth rate of output in marginal firms. Now the latter is not determined by (7), for marginal firms have fixed capacity. Rather the growth rate of output of marginal firms, g_m is determined residually as follows. Let e be the share in aggregate output of profitable firms, i.e. those accumulating according to (7), then $g_D = eg + (1 - e)g_m$ along the balanced path. Taking acount of (8) and remembering that for marginal firms $p = h_m$, we find that $(1 - e)kg_m = kg_D - e[h_m - \bar{h}]$, that is, g_m is determined residually. Whenever $g_D > g$, it will follow that g_m is negative and e is tending towards unity. The practical import of this is that there will be phases of the selection process when $\bar{h}$ is falling while p is constant because it is tied to unit costs of marginal technologies. Once the output of these firms has fallen so they are out of business (that is to zero output given constant returns to scale) then p will again fall at a rate determined by $\bar{h}$ until the next technology and associated firms are pushed into marginality. If $g_D = g$ then $g_m = 0$, while g_D cannot exceed g for long without marginal firms ceasing to be marginal. In the text we ignore this complication.

nal to the variance of unit costs in the industry. In diversity lies progress[20].

The question naturally arises of the long run state towards which this selection process is tending. Clearly it is one in which the dominant firms are the ones which sustain growth longest. and these must be the best practice firms with unit costs indicated by $h_a - h_a$ in Figure 2. Selection draws forth a dominant design from among the set of available technologies. It follows that average practice unit costs will converge upon best practice costs, and that the effect of competition is to eliminate the very diversity on which the strength of selection depends[21].

In short, this selection process is a multi technology diffusion process in which, starting from any initial pattern of market shares, competition selects or draws to market dominance a particular innovation[22]. Selection is thus based upon the twin pillars of diversity in technical knowledge and a positive propensity to accumulate[23].

(ii) A Caveat

The selection process described above depends upon the postulate of a positive propensity to accumulate which is identical for all firms. Now one of the most obvious sources of inter-firm diversity relates precisely to the propensity to accumulate, which moreover, may vary over time in response

20. The link with genetic theory is clear if we identify the growth rate of a firm with the corresponding concept of genetic fitness. Indeed (10) is exactly analogous to R.A. Fisher's Fundamental Law of Natural Selection in Genetic Theory. Cf. R. Nelson and S. Winter, 1983, *An Evolutionary Theory of Economic Change,* Harvard, p. 243. Important precursors of this selection argument are J. Steindl, 1952, *Maturity and Stagnation in American Capitalism,* Blackwell, and J. Downie, 1955, *The Competitive Process,* Duckworth. The process of market share variation is called the "transfer process" in this later book.

21. Note that the equilibrium price would then exceed h_a by the magnitutes kg_D. Unless g_D is zero this would imply a margin of viability above h_a, and indeed this is so. But since any such firms are less efficient than average, it follows that their market shares tend to zero, and they have zero economic weight in determining $\bar{h}$. The distinction between a viable firm and a firm with positive economic weight is an important one in the context of selection and competition.

22. Or group of innovations within the design configuration if they have equal unit costs at the ruling input prices.

23. One measure of the intensity of selection is provided by the variance in the growth rates of the profitable firms. It is readily shown that $V(g, t) = (1/k^2) \, V(h, t) = -(1/k) \, (d\bar{h}/dt)$. Clearly on this measure, the intensity of competition declines in proportion to the decline in average practice unit costs.

to the performance of the firm and the pressures of the competitive environment[24].

The importance of our simplifying postulate may be made clear in the following terms. Notice first, that if all propensities to accumulate are zero it is obvious that the selection process cannot operate in any form. Suppose next, that only one propensity to accumulate is positive, that this is associated with the firms which operate the worst practice technology, and that the growth rate of demand is positive. Then the technology which comes to dominate is precisely the *worst practice technology* since only those firms will be increasing market share. All remaining technologies will remain viable but their economic weight will gradually tend to zero.

To generalize, in a world of divergent propensities to accumulate, two sets of conditions are sufficient to ensure that competition converges upon the best practice process. Either, the best practice firms have a positive propensity to accumulate and the growth rate of demand is zero; or, if the growth of demand is positive, the best practice firms have a propensity to accumulate at least as great as that for any other surviving firm. Failing these two conditions the level of average practice will not converge upon the best practice technology identified in Figure 2[25]. In sum, survival of the fittest cannot imply that the fittest champion best practice technologies, unless there is the appropriate correlation across surviving firms between efficiency and the propensity to accumulate. Unfortunately, space prevents treatment of this topic at any further length.

(iii) The Combination of Product and Process Innovations

Returning to our main theme of technological competition we now allow the innovations to have different product characteristics as well as different process characteristics. We maintain the assumption of a common propensity to accumulate. Introducing product differentiation raises many potential complications but we may avoid most of them by resorting to what is known

24. Compare Marshall's allusions to the life cycle of firms. Cf., G. Shove, 1943, "The Place of Marshall's Principles in the Development of Economic Theory", *Economic Journal,* vol. 53, pp. 294-329.

25. To my knowledge, the point that competition does not necessarily ensure convergence to the best practice technology was first made by S. Winter, 1964, "Economic Natural Selection and the Theory of the Firm", Yale Economic Essays, vol. 4. This valuable paper contains many insights into the working of selection process. On the theme of propensities to accumulate see also E.T. Penrose, 1952, "Biological Analogies in the Theory of the Firm", *American Economic Review*, vol. 42, pp. 804-819.

as the hedonic postulate. This means that each innovation is viewed from the product side as embodying a particular bundle of characteristics (the characteristics being shared by the other innovations in the design configuration) and that these characteristics are valued implicitly by consumers. If v_j is the implicit valuation of characteristic j, then the hedonic postulate permits us to write the price of commodity i as $p_i = \sum_j c_{ij} v_j$, where the c_{ij} are the *known* quantities of characteristics embodied in the product. From this point we may then assign to each innovation a market price premium relative to any "basic" innovation within the design configuration, the choice of which is arbitrary. Given that all consumers have the same marginal valuations for each characteristic, that is, consumption is efficient, these hedonic premia are such that users are indifferent between the various innovations within the configuration. Thus the elasticity of demand for each individual innovation is infinite at a given set of v_j. If p_0 is the price of the basic innovation, we have $p_i = a_i p_0$ with $a_i > 1$[26]. Let us now see how product diversity affects the selection process given a set of fixed characteristic valuations and input prices.

The growth rate of capacity for the ith innovation still depends upon its profitability, but now $g_i = [a_i p_0 - h_i]/h$. For the firms which are profitable, we may then aggregate the various outputs using the fixed set of hedonic premia to give

$$g = \Sigma s_i g_i = \frac{1}{k} [\bar{a} p_0 - \bar{h}] \tag{12}$$

when $\bar{h}$ is again average practice unit cost and $\bar{a} = \Sigma s_i a_i$, is average practice product quality. The selection process translates differential profitability into differential market growth, and so the market shares for each profitable innovation evolve according to

$$\frac{ds_i}{dt} = \frac{1}{k} s_i [p_0(a_i - \bar{a}) + (\bar{h} - h_i)] \tag{13}$$

which identifies above average practice product quality and below average practice unit cost as the twin sources of differential competition advantage.

26. $a_i = \Sigma c_{ij} v_j / \Sigma c_{0j} v_j$. Note that when the hedonic postulate is not applicable we have a complex set of interacting diffusion processes to define selection. Cf. J.S. Metcalfe and M. Gibbons, 1983, "On the Economics of Structural Change and the Evolution of Technology", forthcoming in Proceedings of 7th International Economic Association, Conference, Madrid 1983.

The values of average practice performance will themselves vary over time
as the dominant innovations gain market share and one may then show that
the levels of average practice evolve according to

$$\frac{d\bar{h}}{dt} = \frac{1}{k} \ [p_0 C(a, h, t) - V(h, t)] \tag{14}$$

$$\frac{d\bar{a}}{dt} = \frac{1}{k} \ [p_0 V(a, t) - C(a, h, t)] \tag{15}$$

where $V(a, t)$ is the variance in a across profitable firms, and $C(a, h, t)$ is
the corresponding covariance across the set of a_i, k_i. Notice that these variance and covariance terms are time dependent. Provided that $C(a, h, t) \leqq 0$,
the most process-efficient firms also tend to produce the better quality
products, then $d\bar{h}/dt < 0$ and $d\bar{a}/dt > 0$ — competition produces improvement. However, only when the covariance term is zero will propositions analogous to the Fundamental Law of Natural Selection govern the time paths of
average practice performance. In principle, we have no grounds for ruling
out a positive covariance term, whence it is quite possible to have phases
during the selection process when one, if not both, of the average performance
characteristics is moving towards inferior practice[27].

The argument thus far can be clarified by reference to Figure 3, which
is drawn on the assumption of a fixed set of input prices and user valuations
of characteristics. Values of a_i and h_i associated with the given innovations
in the design configuration are measured in the axes, with the convex area
$b - b$ containing all the relevant innovations, as in Figure 2. On the horizontal
axis to the left of the origin we measure values of kg_D. The ruling value
is given by point G.

At the moment our analysis begins, average practice performance is locat-
ed at point e, corresponding to $\bar{h}(0)$ and $\bar{a}(0)$. Given the growth rate, the only
value of p_0 (and thus the entire price set) which can finance this growth rate

27. Notice that the above depends not upon any particular theory of quality premia
but simply on those premia being given data. K. Cowling and A.J. Rayner, 1970, "Price,
Quality and Market Share", *Journal of Political Economy*, vol. 84, pp. 1292-1309, draw
attention to the role of inertia, ignorance and goodwill in permitting p_i to deviate from the
hedonic premium levels. Let, for example, $p_i = a_i p_0 + u_i$, with u_i a random disturbance
term. Then to (14) we must add the term $(1/k) \ C(h, u, t)$ and to (15) the term $(1/k) \ C(a, u, t)$.
Unless one has any clear *a priori* expectation about the correlation of ignorance, inertia
and goodwill with a_i and h_i, these covariances may perhaps be set equal to zero.

is indicated by the slope of the line $G - G$ (this represents equation 12). Draw OU parallel to G-G and we may then identify the following firm states. Firms in A are bankrupt ($a_i p_0 < h_i$) and those on OU are marginal but not growing. Firms in B are profitable and are expanding absolutely but not relatively, while firms distributed along G-G are dynamically representative. The firms and innovations which are increasing in market share are those located in region C. It will be noted that a firm with below average practice process efficiency may still increase market share if it is of above average practice in the quality of its output.

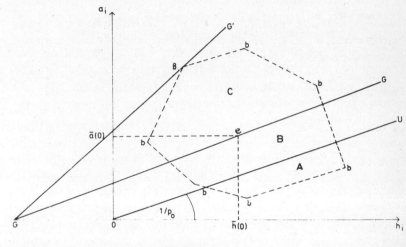

FIGURE 3

The process of selection according to (14) and (15) then converges upon the dominant innovation β, the slope of the line G-G' indicating the terminal basic innovation price. Notice that the best practice innovation does not minimize unit costs but rather minimizes the costs of producing the dominant bundle of product characteristics. As with our process characteristics case, selection equilibrium at a positive growth rate ensures viability for a subset of the initial technologies but economic weight is reserved only for the best practice or dominant design. Of course the caveat raised previously concerning propensities to accumulate applies equally to this more complex case.

D. DIFFUSION AND SELECTION COMBINED

The selection processes described above have been constrained for expositional purposes by two important assumptions. First, the set of innovations is given from the outset with neither improvements within innovations nor imitative behaviour allowed to occur. The economics of competition and technological differentiation could then be made clear. Second, the environment within which selection took place was given, in the sense that we arbitrarily fixed the growth rate of demand, and the structures of input prices and user characteristics valuations independently of the selection process. It is with this assumption of a fixed selection environment that we are concerned in this section, for our treatment of the diffusion process has indicated how the selection environment will change over time as the design configuration is absorbed into its economic environment. Diffusion and selection interact in terms of two competitive processes, that between design configurations and that within design configurations.

As a design configuration is absorbed into its niche, so the selection environment will change in the following general ways. The balanced rate of growth will not remain constant but will typically decline. As the aggregate output of the innovations within the configuration increases, so the price of inputs will change according to their various elasticities of supply and the valuation of characteristics will change according to their elasticities of demand. These induced changes will naturally influence the structures of unit costs and hedonic premia, to continually redefine the innovation which constitutes best practice and alter the evolution of average practice performance. Different diffusion environments will result in the emergence of different dominant innovations, while the innovations, and firms which appear most promising at one stage in the diffusion process may be driven into non-viability by subsequent changes in their environment.

The linking together of diffusion and selection can be illustrated most effectively if we return to the case in which innovations within the design configuration differ only with respect to their process characteristics. If the unit cost function for each innovation is $h_i = h_{0i} + h_{1i}x$, then defining $\bar{h}_0 = \Sigma s_i h_{0i}$ and $\bar{h}_1 = \Sigma s_i h_{1i}$, we have as analogues to (4) and (5)

$$g(t) = \frac{\beta(1 + a\bar{h}_1)}{1 + a\beta k} \left[\frac{c - a\bar{h}_0}{1 + a\bar{h}_1} - x(t) \right] \tag{16}$$

and

$$p(t) = \frac{k\beta c + \bar{h}_0}{1 + a\beta k} + \left[\frac{\bar{h}_1 - k\beta}{1 + a\beta h} \right] x(t) \tag{17}$$

The clear implication is that selection creates drift in the diffusion coefficients, and has the effect of raising the balanced growth rate and lowering the product price at each stage on the balanced path. The curve of diffusion is no longer a logistic curve, although it remains a sigmoid curve.

Introducing different product characteristics greatly complicates the process but leaves the central conclusion intact. The increasing market share of dominant innovations, combined with the elimination of non-viable innovations, induces selective drift in the diffusion coefficients, while the corresponding changes in input prices and hedonic premia shape the rate and direction of selection[28]. The general rules here are as follows.

A decrease in the implicit consumer valuation of a particular characteristic will reduce, relatively, the hedonic price of these innovations embodying relatively more of that characteristic. Correspondingly the profit margins and growth rates of the associated firms will be reduced and capacity of those innovations to acquire an increased market share will be diminished. The environment will have become more hostile towards those particular innovations. Similarly, an increase in the price of a particular input will shift the balance of advantage in gaining market share in favour of the innovations which are least intensive in the use of that input[29]. Naturally these selection induced changes will reduce the growth of demand for inputs which are becoming more expensive and reduce the growth in the supply of characteristics which are becoming less valuable. In turn these changes react back upon the diffusion process, so that diffusion and selection become inseperable.

There remains one further point to stress. We emphasized in our discussion of diffusion the initial uncertainty among potential users as to the merits of any innovation. With respect to selection the role of uncertainty may be developed in two directions. First, with respect to hedonic premia, these will be intimately connected with user appraisal and judgement about the inherently uncertain characteristics contained in the innovations. In effect the set of hedonic premia are likely to change during diffusion, not simply because of changes in user valuations of characteristics, but because of a gradual increase in understanding about the relative technological merits

---

28. These conclusions apply to the balanced path. Deviations from balance will affect the pattern of selection, but not its final outcome. Such developments are beyond the scope of these notes. When h_i depends upon the scale of industry output, the formula for rate of change in $\overline{h}$ becomes

$$\frac{d\overline{h}}{dt} = -\frac{1}{k} \left[V(h_0, t) + x^2 V(h_1, t) + 2 x C(h_0, h_1, t) \right] + \overline{h}_1 \frac{dx}{dt}$$

unless $C(h_0, h_1, t)$ is negative, the bracketed selection effect always works to reduce average practice unit costs.

29. For a brief, formal, treatment see Nelson and Winter, *op. cit.* 243-244.

of different innovations. Secondly, with respect to each firm's uncertainty about its own production process, which will affect both its unit costs and its product quality. Both aspects of uncertainty are likely to be important and will appear again in our comments on learning phenomena.

E. THE INDUCEMENT MECHANISM

We are at last in a position to remove the penumbra of static assumptions surrounding the design configuration, and to link its development with the most important dimension of competition. For it is by active, innovative competition that firms seek to gain product and process advantages relative to their rivals and thus enhance their market share over time. In general, the set of innovations, within the design configuration and the performance of each individual innovation, are not determined from the outset but rather evolve over time as diffusion and selection proceed. The temporal pattern of technical development is a vital element in our understanding of the diffusion and selection processes. For what is being diffused and selected between is a set of co-evolving innovations.

It is worthwhile then to elaborate upon the mechanisms of technical evolution and their interaction with the mechanisms of diffusion and selection.

The timing of market introduction, and the pattern of post innovative improvement of each innovation in the design configuration will be dependent upon decisions made in the relevant firms and, even leaving aside luck and chance, one would expect firms to differ in the propensity to innovate just as they differ with respect to static efficiency and propensity to accumulate. As before the key is to recognize the inherent diversity of the phenomena under investigation.

When considering the propensity to innovate, it is needful to distinguish access to the means to innovate from the incidence of incentives to innovate. With respect to the former, we must distinguish those advances in technology which depend upon the prior commitment of resources (investments in R & D), from those advances in technology which arise as a joint product with the activities of producing and marketing the new commodity, a category which subsumes a variety of learning phenomena.

Consider first the relation between diffusion, selection and learning, and imagine that accumulated experience is specific to the firms in which it is generated. On the process side, learning will depend upon accumulated production experience, while on the product side, it will depend upon accumulated user experience and the communication of this to the firms in question. Interaction between innovators and users has rightly been identified as a key element in the innovation process. Its importance arises largely because the markets for product characteristics are implicit so that market research and producer/user interaction have to substitute for well defined price signals[30].

Given a firm's capacity to learn — which may not be equal in all directions — diffusion and selection determine which firms gain experience most rapidly. The more efficient firms grow more rapidly and experience the greatest reductions in unit costs and improvements in product quality, so augmenting their growth in a virtuous circle of advance. Average practice costs fall more rapidly and diffusion proceeds more quickly relative to the no learning situation. Not all learning economies need be firm specific, and to the extent that they may be imitated by other firms, this further accelerates the improvement of average practice. However, one expects the scope for learning to be limited and so the incidence of learning effects is likely to be greatest in the early stages of diffusion.

We turn now to those advances in technology which require a prior commitment of research and development resources. The first point on which to be clear is that in general the resources to undertake commercial R & D must be financed out of the same profit streams which support the accumulation of capacity. Firms have a choice between innovation and accumulation and the way in which they exercise this choice is crucial to their long-term viability. Given this choice the capacity of a firm to innovate will depend on its market share and the aggregate flow of industry profits. As diffusion and selection interact so the distribution of resources to innovate will change. In particular, the selection mechanism will continually redistribute profits toward the firms of above average productive efficiency and product quality, and provide them with the means to further enhance their performance. Again we find a virtuous circle of interdependence between efficiency, quality, market share and innovation related increases in efficiency. Furthermore this mechanism operates in the context of the diffusion induced decline in the rate of profits as the industry approaches its equilibrium niche. The middle part of the diffusion path is then likely to provide the greatest aggregate profits to support innovation, as Figure 1d illustrates. It is not only the resources for innovation which are relevant to this discussion but also the costs of advancing technology. Here several elements need to be incorporated into the picture. On the one hand, firms may differ in the efficiency with which they manage the research and development process. Static efficiency need not imply efficiency at innovation. On the other hand, the

30. On "learning by using", see the insightful paper by N. Rosenberg op. cit.

costs of incremental advance may differ significantly between the rival process technologies and between product and process improvements. For any one process we might expect that the marginal cost of advance at first falls, reflecting the cumulative way in which knowledge builds upon knowledge, but then, as the limits to the potential of the technology are approached, the marginal costs of advance rise sharply. Rising marginal costs of R & D and depleted profits jointly imply a slackening of the rate of progress.

Having discussed the capacity to innovate, it remains to make the connection with the incentives to innovate, and it is necessary here to distinguish incentives associated with the scale of output, from incentives associated with the pattern of input prices and user valuations of characteristics. It has long been recognized that R & D is an increasing returns activity in the sense that the cost of advancing technology is independent of the scale at which advances are applied to production[31]. Hence the economic return to a post-innovative improvement rises in proportion to the scale at which it is applied. Within the present context several implications follow from this. First, the growth of the industry as a whole increases the general incentive to conduct R & D. Secondly, the selection process continually redistributes this incentive in favour of firms with increasing market shares, precisely the firms which are already of above average performance[32]. As diffusion and selection interact, the resources and incentives to advance technology become progressively concentrated on the technologies operated by a decreasing number of expanding firms. Concentration and the evolution of technology proceed in step.

We are thus led to stress the interdependence between the diffusion, selection and inducement mechanisms. The growth of industry output and its concentration in progressively fewer firms increases the incentives to innovate. Technical progress might then continue indefinitely but for the limited potential of each individual process technology. It is to the rising incremental costs of advance that we must look to retard and ultimately terminate the rate of progress, and normally this will imply an end to progress before the ultimate performance limits of a technology are fully explored. Technological maturity is in the end an economic concept. The slackening of technical progress contributes to the squeeze upon profit margins and

31. Cf., J. Schmookler, 1966, *Invention and Economic Growth*, Harvard, also more recently, P. Dasgupta and J. Stiglitz, 1980, "Industrial Structure and the Nature of R & D Activity", *Economic Journal*, vol. 90.

32. One might argue that the incentive depends on expected market share rather than, myopically, on the prevailing market share. While formally correct it should be remembered that the costs of gaining a higher market share must then added to the R & D costs.

in turn this results in further resource and market limitations on the pace
of progress. The absorption of the technology into its long-run niche and the
maturing of the technology are jointly determined aspects of one and the
same process. Not only is progress retarded in the larger firms, the rising
cost of advance takes innovation beyond the resource limits of firms with
smaller market shares and they may cease to be technically progressive long
before they cease to be viable. At best the ability of such a firm to survive rests
on its capacity to imitate. The nature of property rights in technology and
the ease with which technology can be transferred within the industry, either
through imitation or licencing, become crucial to the survival of these un-
progressive firms. It should always be remembered that the incentives
to innovate have to be matched by the resources to innovate, and that selection often results in firms with the incentives but without the resources. In
sum, the selection mechanism works not only on technologies but on patterns
of technical advance.

The second dimension of the inducement mechanism, that relating to
relative prices, draws upon the familiar idea that different patterns of input
prices will result in different patterns of post innovative improvement, dependent upon the balancing of changed economic circumstances and the costs
of different directions of advance[33]. Similarly, different implicit user valuations of characteristics will bias the profitable directions of product improvement. In each case the pattern of technical change will evolve to economize
on inputs which are becoming relatively more expensive and to increase the
relative importance of characteristics which are becoming more valuable.

In terms of the selection process the effect will generally be to increase
the rate at which average practice cost declines, and to increase the rate at
which average product performance is improved[34]. In terms of the diffusion

33. After a shaky start this proposition now has a firm foundation. See, in particular,
S. Ahmad, 1966, "On the Theory of Induced Innovation", *Economic Journal*, vol. 76,
and H. Binswanger, 1974, "A Micro Economic Approach to Induced Innovation", *Economic Journal*, vol. 84. Strictly speaking the proposition should relate to the expected future pattern of relative input prices.

34. Return to the simplest case of process differentiation, and costs independent of
industry output. Let $\lambda_i(t)$ be the (neutral) rate of reduction of unit cost for the ith innovation so that $dh_i/dt = -\lambda_i(t)h_i$. It then follows that

$$\frac{d\bar{h}}{dt} = -\frac{1}{k}[V(h, t) + C(\lambda, h, t) + \bar{\lambda}\bar{h}], \text{ and that } \frac{d\bar{\lambda}}{dt} = -\frac{1}{k}C(\lambda, h, t)$$

Again the sign of the covariance is important to the outcome, but $C(\lambda, h, t) \geqslant 0$ guarantees
a more rapid rate of decline in h than would otherwise occur.

mechanism, induced progress will relax the market and resource constraints on growth and accelerate the general rate of diffusion. Thus as the design configuration is integrated into the economic structure we expect to observe the emergence of well defined trajectories of technical development, and it seems clear that different diffusion patterns will produce different patterns of technical progress. As a minimum it would appear essential to locate the design configuration within its network of inter-industry input-output relationships, before the full ramifications of the inducement mechanism can be explored.

By Way of Synthesis

We may now draw together the various threads of this analysis. Starting from the proposition that any configuration has a post-innovative potential for improvement we have sought to determine its trajectory of evolution in terms of the incentives and opportunities facing the firms "championing" each component technology. Chance apart, the pattern of evolution will be increasingly dependent on the activities of a declining number of producers with the resources and the market incentive to advance their particular technology.

Patterns of technical change and patterns of diffusion are simultaneously determined, and recognition of this leads to important insights into the diffusion process. In particular what is typically diffused is not a given innovation but a set of innovations whose technologies are changing systematically in response to experience and the incentives arising during diffusion. Such technical progress accelerates the rate of decline of average practice unit costs and thus increases the pace of diffusion. If progress is fast enough we may expect an acceleration of the growth rate of the industry, until the rising marginal costs of innovation begin to guide the industry towards its long-period niche. Clearly the long-period position is subject to continuous redefinition as technologies evolve. All this takes place in the context of selection between the different trajectories of technical advance posed by rival innovations. An innovation which is initially relatively inefficient may yet evolve into the best practice technology, if its potential is sufficient, if it is championed by progressive firms and if the movement of relative input prices and user valuations is favourable to its dominance. Sufficient qualification to ensure that different diffusion and selection environments could produce quite different dominant technologies.

REFERENCES

AHMAD, S. (1966), "On the Theory of Induced Innovation", *Economic Journal*, vol. 76.

ANDREWS, P.W.S. (1964), *On Competition in Economic Theory*, Macmillan.

BINSWANGER, H. (1974), "A Micro Economic Approach to Induced Innovation", *Economic Journal*, vol. 84.

CARTER, C.F. and WILLIAMS, B.R. (1959), *Industry and Technical Progress*, Oxford University Press.

COWLING, K. and RAYNER, A.J. (1970), "Price, Quality and Market Share", *Journal of Political Economy*, vol. 84, pp. 1292-1309.

DASGUPTA, P. and STIGLITZ, J. (1980), "Industrial Structure and the Nature of R & D Activity", *Economic Journal*, vol. 90.

DAVIES, S. (1979), *The Diffusion of Process Innovations*, Cambridge University Press.

DOWNIE, J. (1955), *The Competitive Process*, Duckworth.

FREEMAN, C. (1984), *Long Waves in the World Economy*, Butterworth.

GOLD, B. (1981), "Technological Diffusion in Industry : Research Needs and Shortcomings", *Journal of Industrial Economics*, vol. 24.

HAYEK, F.A. (1964), *Individualism and Economic Order*, Chicago.

HICKS, J. (1977), *Economic Perspectives*, Oxford University Press, p.p. 16-17.

HOFFMAN, W.G. (1.949), "The Growth of Industrial Production in Great Britain", *Economic History Review*, vol. 4.

IRONMONGER, D. (1972), *New Commodities and Consumer Behaviour*, Cambridge University Press.

KALDOR, N. (1934), "The Equilibrium of the Firm", *Economic Journal*, vol. 44, pp. 60-76.

LANCASTER, K.J. (1971), *Consumer Demand : A New Approach*, Colombia.

LANGRISH, J. (et al.) (1969), *Wealth from Knowledge*, Macmillan.

McNULTY, P.J. (1968), "Economic Theory and the Meaning of Competition", *Quarterly Journal of Economis*, vol. 82, pp. 639-656.

METCALFE, J.S. (1981), "Impulse and Diffusion in the Study of Technical Change", *Futures*, vol. 13.

—— and GIBBONS, M. (1983), "On the Economics of Structural Change and the Evolution of Technology", Proceedings of 7th International Economic Association Conference, Madrid.

NELSON, R.R. (1981), "Research on Productivity Growth and Productivity Differences : Dead Ends or New Departures", *Journal of Economic Literature*, vol. 19.

—— and WINTER, S. (1983), *An Evolutionary Theory of Economic Change*, Harvard, pp. 243-244.

PARKINSON, C.N. (1966), *A Law Unto Themselves*, John Murray.

PASINETTI, L. (1981), *Structural Change and Economic Growth*, Cambridge University Press, pp. 75-76.

PENROSE, E.T. (1952), "Biological Analogies in the Theory of the Firm" *American Economic Review*, vol. 42, pp. 804-819.

—— (1961), *The Theory of the Growth of the Firm*, Blackwell.

ROSENBERG, N. (1982), *Inside the Black Box : Technology and Economics*, Cambridge University Press.

ROTHWELL, R. and ZEGVELD, W. (1982), *Innovation and the Small and Medium Sized Firm*, Frances Pinter.

SALTER, W.E.G. (1961), *Productivity and Technical Change*, Cambridge University Press.

SCHMOOKLER, J. (1966), *Invention and Economic Growth*, Harvard.

SCHUMPETER, J. (1934), *The Theory of Economic Development*, Oxford University Press.

—— (1947), *Capitalism, Socialism and Democracy*, George Allen and Unwin.

SHOVE, G. (1943), "The Place of Marshall's Principles in the Development of Economic Theory", *Economic Journal*, vol. 53, pp. 294-329.

STEINDL, J. (1952), *Maturity and Stagnation in American Capitalism*, Blackwell.

STONEMAN, P. (1983), *The Economic Analysis of Technological Change*, Oxford University Press.

WILLIAMS, B.R. (1949), "Types of Competition and the Theory of Employment", *Oxford Economic Papers*, vol. 1.

WINTER, S. (1964), "Economic Natural Selection and the Theory of the Firm", *Yale Economic Essays*, vol. 4.

CHAPTER THREE

REINDUSTRIALISATION, INNOVATION AND PUBLIC POLICY

By Roy Rothwell*

Introduction

There is a growing belief among governments in the advanced market economies that one means of at least partially overcoming the current world economic crisis is the stimulation of radical industrial innovations. Indeed, there exists a great deal of evidence to suggest that the roots of the current world economic crisis go back further than the first oil crisis of 1973 and are very much bound to the rate and nature of technological innovation (Mensch, 1979; Rothwell, 1981; Freeman, Clark and Soete, 1982). In the late 1960s a large number of the "new" industries of the post-war period entered more or less simultaneously the maturity and market saturation phase of their life cycles in which the technological emphasis largely was on manufacturing process change and rationalisation. At the same time there was a paucity of radical product innovations to regenerate market demand and to create new techno/economic combinations. As a result productivity increase — albeit at a generally reduced rate — began to outstrip output (demand) growth and many manufacturing jobs were shed. The oil crisis significantly exacerbated this already established trend towards stagnation.

More recently we can point to the emergence of a set of potential new techno/economic combinations on which the new industries of the future will be based. At the same time, a number of these new technologies have great potential for the regeneration of existing sectors through both major product improvement and dramatic productivity increase (e.g. information technology, robotics, biotechnology). The implication of this "structural" interpretation of the present crisis is that it should not be tackled by traditional demand management policies, but by technology-oriented reindustrialisation policies. To be sure, while an adequate macro-economic stabilisation policy is necessary to create a climate favourable to innovation and reindustrialisation it is, by itself, insufficient to induce the necessary radical innovative activity.

For the purpose of this paper, reindustrialisation can be defined as :

* Science Policy Research Unit, University of Sussex. This article is based on Rothwell and Zegveld (1981, 1985 forthcoming).

"The structural transformation of industry into higher value added, more knowledge - intensive sectors and product groups, and the creation of major new technology-based sectors and products serving new markets".
(Rothwell and Zegveld, 1985 forthcoming, Chapter 1)

During the post second world war period we have seen a number of examples of reindustrialisation, two of the most dramatic having taken place in the United States and Japan. In the first case, companies in the United States were instrumental in the creation of a set of new techno/economic combinations resulting in the rapid growth of a number of new high technology industrial sectors, for example the semiconductor, computer aided design (CAD) and satellite communication industries. This resulted in a shift in the balance of US industrial production and exports to more technology-intensive product groups (Aho and Rosen, 1980). While public (mainly defence) procurement and government R & D funding played a crucial role in the United States, the reindustrialisation process was not the result of overt (formal) public reindustrialisation policies, although it has been suggested that the various US administrations did effectively operate covert (informal) policies via military and space R & D funding and equipment procurement programmes (Rothwell and Zegveld, 1985 forthcoming, Chapter 4).

In the case of Japan, industrial transformation was the result of deliberate public policy. Following the second world war, the decision was made in Japan deliberately to restructure its industry better to match the evolving market and technological requirements of the third quarter of the twentieth century. This process has been described by Allen (1982) :

"A corollary of Japan's determination to lift her productive capacity to a higher plane of technical and commercial competence was her awareness of the need to adapt her industrial composition to changes in markets and techniques. Structural adaptability was recognised as a condition of the continuous growth in GNP. Moreover, Japan did not simply respond to exogenous forces; she was remarkably successful in anticipating change In the early 1950s her chief export industries still consisted of labour-intensive trades, where her low wages made her an effective competitor with her western rivals, and her superior management and organisation kept at bay challenges from the developing countries. But, as her policy showed, she fully realised these advantages were transient, and she soon began to set up a new capacity in several large scale, capital intensive trades, notably steel, shipbuilding and chemical

fertilizers. By the early 1960s, when these trades were well established, she turned her attention to a number of engineering industries, especially radio, television sets and motor vehicles, as well as to petrochemical products. By the end of the 1960s her motor car, electronics and watch and clock industries ranked with the world's leaders, while her eminence in steel, motor cycles and shipbuilding remained unassailed. The check to her industrial growth after 1973 was followed by a recovery (in 1978) which was associated especially with the further development of her motor, machinery, instrument and electronic manufactures".
(Allen, 1981, pages 69 and 70)

This structural transformation of Japanese industry can be depicted in Figure 1, which is a chart used the Japanese Economic Planning Agency to explain Japan's economic development. It well illustrates Japanese movement towards higher value added, more knowledge intensive sectors. Moreover, unlike the United States where reindustrialisation involved the initiation of *new* techno/economic combinations, in Japan reindustrialisation was based on the acquisition and subsequent improvement of *existing* technological know-how (mainly from the United States); in other words, Japanese re-industrialisation was based on a process of rapid technological "catching up".

Innovation Policy

While in the final analysis a nation's ability to undergo structural indu-strial transformation will depend to a large extent on the abilities and pro-pensities of industrial managers, there seems little doubt that public policies also have an important role to play in this process. Public policies can en-hance the technological potential both of individual companies and public R & D institutions; they can remove perceived barriers to firm-based innovations; they can promote an overall environment conducive to firm-based investment in techno/market activities and public bodies can create an innovation-demanding market through their procurement activities.

Policies for promoting scientific and technological advance have, of course, been in existence for many years. They include the patent system, technical education and the funding of basic and applied research within the scientific and technological infrastructure. Similarly industrial policies are of long standing and include such measures as taxation policy, investment grants, tariff policy and industrial restructuring. More recently, a number of governments have become involved in the formulation of explicit *innova-tion* policies. If we define technological innovation to include "the technical,

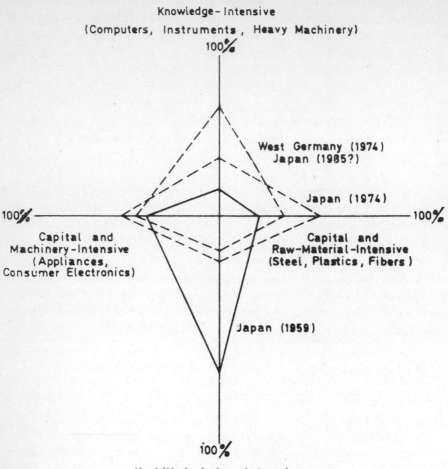

FIGURE 1
The Evolution of Industial Structure

financial, manufacturing, management and marketing activities involved in the commercial introduction of a new product, or in the first commercial use of a new manufacturing process or equipment", then we can see that innovation policy is essentially a fusion of the longer-standing science and technology and industrial policies. Innovation policy includes the whole sequence of activities involved in the commercial introduction of a new product or process, from basic research through to marketing and sales. Thus, innovation policy is essentially an integrative concept.

Having defined innovation policy, we can go on to list a set of innovation policy tools — see Table 1. These tools can be divided into three broad categories. These are :

(i) Supply side tools   — these include the provision of financial and technical assistance, including the establishment of a scientific and technological infrastructure.

TABLE 1

Classification of Government Policy Tools*

Policy tool	Examples
1.– Public enterprise	Innovation by publicly owned industries, setting up of new industries, pioneering use of new techniques by public corporations, participation in private enterprise
2.– Scientific and technical	Research laboratories, support for research associations, learned societies, professional associations, research grants
3.– Education	General education, universities, technical education, apprenticeship schemes, continuing and further education, retraining
4.– Information	Information networks and centres, libraries, advisory and consultancy services, detabases, liaison services
5.– Financial	Grants, loans, subsidies, financial sharing arrangements, provision of equipment, building or services, loan guarantees, export credits, etc.
6.– Taxation	Company, personal, indirect and payroll taxation, tax allowances
7. –Legal and regulatory	Patents, environmental and health regulations, inspectorates, monopoly regulations
8.– Political	Planning, regional policies, honours or awards for innovation, encouragement of mergers or joint consortia, public consultation
9. –Procurement	Central or local government purchases and contracts, public corporations, R & D contracts, prototype purchases
10.– Public services	Purchases, maintenance, supervision and innovation in health service, public building, construction, transport, telecommunications
11.– Commercial	Trade agreements, tariffs, currency regulations
12.– Overseas agent	Defence sales organisations

* This table is based on Table 1 in Braun, (1980).

Source : Rothwell and Zegveld, (1981), Industrial Innovation and Public Policy, London Frances Pinter.

(ii) Demand side tools   — these include central and local government purchases and contracts, notably for innovative products, processes and services.

(iii) Environmental tools   — these include taxation policy, patent policy and regulations (worker health and safety, environmental and economic).

An analysis several years ago of public innovation policy statements in six countries, which looked at the various policy recommendations by type of tool, highlighted a number of interesting differences between nations — see Table 2.

TABLE 2

Analysis of Policy Recommendations by Type of Tool

Type of tool	Canada	Japan	Netherlands	Sweden	United Kingdom	United States
1. Public Enterprise	0	0	0	1	1	0
2. Scientific and technical	7	7	9	3	4	4
3. Education	3	1	5	11	4	3
4. Information	2	2	8	2	3	8
5. Financial	5	2	6	5	6	4
6. Taxation	1	0	0	1	6	13
7. Legal and Regulatory	0	0	6	1	0	46
8. Political	2	4	2	3	4	2
9. Procurement	1	0	2	2	3	11
10. Public Services	0	0	1	0	3	0
11. Commercial	2	1	1	0	0	3
12. Overseas agent	0	0	0	1	2	0
Total	26	17	40	30	36	94

Source : Rothwell and Zegveld, (1981).

For Canada, Japan and the Netherlands, for example, the most favoured tool is the "scientific and technical" one, for Sweden "education", for Britain "financial and taxation" and for the United States "legal and regulatory". Indeed, something like fifty per cent of tools recommended for the United States are in the "legal and regulatory" category, which reflects a strong belief that US industry is over-regulated (Rothwell, 1980, 1981b, National Academy of Sciences, 1980).

So far we have dealt with innovation policy tools rather than the national philosophies underlying their use, and consideration of national strategies

TABLE 3
Areas of Interest to Japanese Industry

New Products	Energy industries	Advanced, high-technology industries
Optical fibres	Coal liquidation	Ultra-high-speed computers
Ceramics	Coal gasfication	Space developments
Amorphous materials	Nuciser power	Ocean developments
High-efficiency resin	Solar energy	Aircraft
	Deep geothermal generation	

Source : Japanese Ministry of International Trade and industry.

TABLE 4
Strategic High-Technology Priorities in France

Strategic industry	Objectives	Overall actions planned
Electronic office equipment	To achieve 20-25% world market share, and avert an anticipated $ 2bn trade deficit in 1985	In strategic sectors the government will negotiate development contracts with individual companies, setting specific goals for sales exports and jobs. Firms that make such commitments will receive tax incentives, subsidised loans, and other official aids.
Consumer electronics	To create a world-scale group including TV-set and tube makers that will each rank among the top three globally. To eliminate the $ 750 million trade deficit in such products.	
Energy-saving equipment	To ensure that government grants to companies and households to install such equipment are spent primarily on French products.	
Undersea activities	To recapture second place in the world after the USA.	
Bioindustry	Objectives not yet defined	
Industrial robots	Objectives not yet defined.	

These six industries together are expected to add $ 10 bn in sales and to double their world force to 135.000 by 1985.

Source : Business Week, 30 June 1980, page 140.

towards innovation highlights a number of important differences in this respect. Perhaps the most significant difference lies between those nations that have a clear cut, long term strategy towards the development and exploitation of specific high technology product groups and new technologies, and those that do not. In Japan and France, for example, (see Tables 3 and 4) there is a clear emphasis on attempting to identify potentially important new industrial sectors. These nations appear to have accepted that structural change to their economies, via major technological innovations, is to be actively pursued, arguing that greater advantage is to be gained from exploiting changes in the new world economic order rather than steadfastly resisting these changes through measures seeking to protect ailing industries. In other words, they lay emphasis on the structural transformation of the industrial base towards newer, more technology- intensive sectors and product groups; that is, their policies contain a strong "reindustrialisation" component.

Some Difficulties with Innovation Policies

While innovation policies in Japan appear to have been remarkably effective, in most countries they have suffered from a variety of problems. Despite the lack of systematic and objective assessments of public innovation policies, it is nevertheless possible to identify, in a rather general sense, a number of problems innovation policies have in the past suffered from (Rothwell, 1982).

First, there has often been a lack of market know-how amongst public policy makers, and according to Golding (1978) and Little (1974) there exists evidence to suggest that government funds often have gone to support projects of high technical sophistication but of low market potential and profitability. As well as often having lower market potential, projects funded by governments have also tended to involve higher technical and financial risks than those funded wholly by industrial companies. That government backed projects involving high technical risk might, of course, be taken as justification for government involvement in the first place ; the problem governments face is to identify high-risk projects that also have high market potential, yet it is doubtful whether government policy makers generally possess the competence to assess market prospects satisfactorily.

Second, subsidies have in the past tended to assist mainly large firms, an imbalance that can and should be redressed. This tells us nothing, of course, but the effectiveness of those subsidies that have been made to small firms. In the case of large firms, evidence from Canada and the FRG suggests that funds often have gone in support of projects that are relatively small and sometimes of dubious merit such that the companies would not, by themselves, finance them. The Enterprise Development Program in Canada (Rothwell and Zegveld, 1981, p. 199), which will support only projects that represent a significant risk to the firm, represents an explicit attempt to avoid financing marginal projects in large firms.

Third, governments often have tended to adopt a passive, rather than an active, stance towards information dissemination. As a result, policy measures have been taken up largely by a limited number of "aware" (usually large) companies.

Fourth, there has been a general lack of practical knowledge, or imaginative conceptualisation, of the process of industrial innovation by policymakers. As a result, they have tended to adopt an R & D — oriented view of innovation, often to the detriment of other important aspects of the process, e.g. innovation-oriented public purchasing.

Fifth, there has sometimes been a lack of interdepartmental coordination — and sometimes cooperation — between the relevant organisations and agencies involved in the policy process. This can result in lack of complementarity between different initiatives, and might also lead to the propagation of contradictory measures. That this is true in other areas of public policy is amply demonstrated in Table 5 for the US.

TABLE 5

Some Contradictory Policies Affecting Industry in the USA

The Environmental Protection Agency is pushing hard for stringent air pollution controls.	The Energy Department is pushing companies to switch from imported oil to "dirtier" coal.
The National Highway Traffic Safety Administration mandates weight-adding safety equipment for cars.	The Transportation Department is insisting on higher vehicles to conserve gasoline.
The Justice Department offers guidance to companies on complying with the Foreign Corrupt Practices Act.	The Securities and Exchange Commission will not promise immunity from prosecution for practices the Justice Department might permit.
The Energy Department tries to keep down rail rates for hauling coal to encourage plant conversions.	The Transportation Department tries to keep coal-by-rail rates high to bolster the ailing rail industry.
The Environmental Protection Agency restricts use of pesticides.	The Agricultural Department promotes pesticides for agricultural and forestry use
The Occupational Safety and Health Administration chooses the lowest level of exposure to hazardous substances technically feasible, short of bankrupting an industry.	The Environmental Protection Agency uses more flexible standards for comparing risk levels with costs

Source : Business Week, 30 June 1980.

Finally, there has been a tendency for innovation policies to be subjected to changes in political philosophy rather than to changing national or international economic needs or conditions. Thus, in many countries policies have been subjected to a political cycle rather than to the dictates of economic, industrial or technological cycles.

In a survey seeking to determine the instrumentality of public technological innovation policies Goldberg (1981) showed there to be a marked paucity of empirical studies of policy effectiveness. This conclusion was supported by Gibbons (1982) who could point to only two major international studies, those of Allen et al. (1978) and Rubenstein et al. (1977). In the former study, while there were some variations between sectors of industry and between countries on the types of tools influencing firms, Allen et al. were forced to conclude that in general

"the most significant aspect of the study lies in its failure to detect any effects on project performance, of government attempts to stimulate innovation... If project success or failure can be taken as any measure of effectiveness of their actions, then little can be said to have resulted from all this /governmental/ expenditure of effort and money".

The results of the second study led the authors to draw essentially similar conclusions, again despite some differences between countries in managers' perceptions of the influence of government support for industrial activity. Thus,

"... there are a number of basic and important perceptions among managers which are very similar across countries and which support our two secondary propositions that (1) government action to stimulate innovation is perceived as comparatively irrelevant and that (2) government actions generally delay the R D/I process. There is the belief that the effect of market forces and competition on the R D/I process outweighs by far the effects of government actions and that general government policies far outweigh the effects of specific inventive programmes. It is only in rare instances that the incentive programmes are perceived to have any direct effect on specific R D/I decision making. These commonly shared perceptions are joined by the feeling that the incentive programmes are in general too inflexible and too demanding in terms of required administrative details and liaison effects. In the administration of incentive programmes, governments are seen to be too slow and complex in their response to the needs of industry".

Despite this rather dismal picture concerning the influence, or lack of it, of government policy on technological change processes in industry, there is one important caveat to bear in mind. That is, both researchers employed only *one* measure of effectiveness, namely the perceptions of industrial managers, and this might be an insufficient guide to reality. Certainly managers' perceptions of the influence of government policy can often be greatly biased, which was evidenced in the results of a Workshop on Government Regulation and Industrial Innovation jointly organised by the Six Countries Programme on Innovation and the US National Science Foundation in 1979. This showed the perceptions by managers in the US that direct regulatory impacts were almost always very large and negative were, with the exception of a few sectors of industry, considerably overexaggerated (Rothwell, 1979). These perceptions were, moreover, largely responsible for the current US trend towards deregulation. There does, in fact, appear to be a general tendency for managers to overestimate the negative influences of governmental intervention and to understate its positive influences.

It is also essential that the outcome of a policy initiative be measured against its original intention. For example, while in some instances government regulations can have significant positive or negative influences on particular innovations or technologies, only seldom are regulations formulated that are designed with a primary purpose explicitly to stimulate commercially oriented industrial technological developments. Thus, only in certain cases can regulatory policy be viewed as a direct and explicit element of technological innovation policy and in general it should not be assessed as such. Explicitly used as an arm of innovation policy, regulatory policy can, however, be a powerful enabling and even stimulating tool and in such instances should be assessed as such.

A more fundamental point, and one made by Gibbons (1982) is the suggestion that the Allen and Rubenstein studies did not *evaluate* policies, but rather they *monitored* them. In other words they were concerned with the question *What happened*" rather than with the question *"What difference has it made"*. This distinction has clear methodological implications, the most significant perhaps being the need for *comparative* evaluations.

A significant fact to emerge from the various innovation studies that have been undertaken is that, outside the United States (where public procurement and government regulations have been important), governments have rarely been seen to play a major determining role in innovatory success or failure at the level of the firm. In some instances this has undoubtedly been a function of the research methodologies employed, which have often concentrated solely on firm-specific factors. It might also point to a more funda-

mental point, and one made earlier, which is that innovatory success or failure, in practice, is determined largely by the actions of management. The
implication of this, of course, is that in the case of specific innovation projects
in firms, there are inherent limitations to what public innovation policies
can achieve, and public policy makers would do well to bear this in mind.
While innovation policies can enhance the performance of competent, technically progressive managers, they can do little in the face of managerial incom-
petence or simple indifference; in other words, innovation policies are no
panacae for success.

Reindustrialisation Policy

According to Rothwell and Zegveld (1985 forthcoming, Chapter 9), from
the viewpoint of reindustrialisation, public policy must simultaneously tackle
three main factors determining overall national innovative performance : technological opportunity; structure of the industrial sector; size and structure
of market demand. Governments must provide a suitable regulatory framework in which all three elements effectively can develop, and the three remaining main government policy instruments — finance, procurement, technical
infrastructure (including technical education) — should be directed at these
elements in a balanced way.

An important factor influencing national technological opportunity is
the size and orientation of the scientific and technological infrastructure
(universities, government laboratories and collective industrial research
institutes). In the case of universities, it is clear that greater attempts should
be made both to orient the structure of university courses better to match the
evolving manpower needs of the market sector and of society at large, and to
improve the linkages between university research and the market sector.
The same can be said of research performed in government laboratories.
In the case of collective industrial research institutes, while they may be reasonably well adapted to assist in the technological transformation of traditional, end of cycle industries (they were generally set up many years ago
to support such industries), their skill profiles generally are locked into
the industrial requirements of the nineteen sixties and fail to match the evolving technological needs of the nineteen eighties, specifically those of start of
cycle industries. Clearly much greater flexibility will be required on the part
of these infrastructural institutions if they are to play a significant part in
the industrial transformations (notably the creation of new techno/economic
combinations) that are the main aim of reindustrialisation policy. In this
respect, perhaps the temporary joint public/private research facilities esta-

blished in Japan for the development of specific high technology product groups (e.g. very large scale integrated circuits, fifth generation computers) provide a useful model (Sigurdson, 1984).

Turning to the structure and dynamics of the industrial sector, evidence from the United States suggests that the emergence of new techno-economic combinations is associated with a process of complementary and dynamic interaction between existing large corporations and (initially) small new technology-based start ups (Rothwell and Zegveld, 1985 forthcoming, Chapter 6). This was demonstrably the case with the semiconductor indusrty (Rothwell and Zegveld, 1982), the computer aided design industry (Kaplinsky, 1982) and is currently the case with the emerging "new wave" biotechnology industry (OTA, 1984). The point is, public policies generally have swung from supporting industrial agglomeration (big is beautiful) to a bias in favour of small firms (small is beautiful), and have largely ignored the dynamic complementarities that exist between the two. While the balance between the large and the small might vary over the industry cycle, it should be a prime aim of public policy to redress any major imbalances that occur.

With respect to the size and structure of market demand :

"The size and structure of the demand side clearly are key elements in determining innovative-performance. Consideration of market structures and dynamics is thus an important element in reindustrialisation and technology policy. Compared with Europe, the USA and Japan both have large internal markets, and ones which adapt easily to, and indeed help to create, a technically sophisticated supply. From the standpoint of achieving international competitiveness, firms in these countries have first to compete successfully in their national markets because of the severe levels of internal competition existing there. In the effectively fragmented markets in Europe, fierce competition on more than a national basis is often lacking, largely because of the existance of many non-tariff barriers. Even within the EEC, many obstacles prevent the functioning of one "common" market; technical and administrative barriers to trade and subsidies to industrial and agricultural firms all distort competition in the internal market. Trade liberalisation, including the harmonisation of standards within Europe, by doing away with non-tariff barriers, is a major avenue for European reindustrialisation policy and would provide a sensible framework for achieving greater competitiveness in world markets". (Rothwell and Zegveld, 1985 forthcoming, Chapter 9).

For a large variety of products governments provide substantial markets and are hence in a position to exercise their market power in influencing the direction of supply towards higher value added, technologically more innovative products. Thus, public procurement policy can be considered, potentially at least, to be an effective instrument to influence both the rate and direction of supplier innovations. Given the fact that in the advanced market economies governments (central and local) purchase probably between 30 per cent and 50 per cent of all goods and services, it is surprising how little apparent awareness there has been on the part of public purchasing agencies of their potential for influencing technological change.

Innovation-oriented procurement policy fits well the requirements of reindustrialisation. The more traditional policies of support, encouragement, experiment and adaptation of new technologies should be complemented by more ambitious long-term, technology-stimulating procurement strategies. It is, however, clear that there is a great deal of tension between the requirements of innovation oriented procurement — a notable feature being the emphasis on performance rather than price factors — and the current policies of many western countries which are primarily directed at reducing government deficits, i.e. which are reducing the market role of governments, at least outside of the defence sector.

Turning now to finance, this is, in a sense, a "generic" element of reindustrialisation policy since it can greatly influence all other main elements. For example, public finance can significantly influence the rate and direction of infrastructural (and firm based) technological change; financial systems (including grants for equipment, innovations etc.) can influence the structure and operations of the industrial sector; and the size and structure of market demand is to an extent influenced by capital availability and public policies affecting interest rates, tax levels, public sector expenditures, and so on. In this respect it is perhaps government financial strategies that act as the main link between technology and reindustrialisation policies and macroeconomic policies.

According to Rothwell and Zegveld (1985 forthcoming) :

"In the generic area of finance, we see three broad levels of policy.

— Finance for R & D : this includes orienting finance of infrastructurally-based R & D towards stimulating developments in main priority areas and in facilitating transfers to industry. It includes also utilising government grants to orienting industrial R & D towards reindustrialisation projects, i.e. achieving complementarity between the industrial and infrastructural streams of technological development.

— Finance and industrial structure : this involves influencing financing
systems (both public and private) towards achieving the appropriate
industrial structural dynamic; in general, it means increasing the
availability of "patient' money for long-term restructuring programmes in firms and of venture capital for new technology-based start ups.

— Overall fiscal climate : this involves establishing an overall climate
conducive to private investment in reindustrialisation projects; favourable tax regimes, directed public expenditures, moderate interest
rates, and so on".

It should be a major aim of reindustrialisation policy to ensure that
financial policies at all three levels complement each other. Perhaps the most
useful example in this respect, of overall policy coherence, can be found in
Japan. Below is presented a rather simple policy "model" that includes government regulations and financial policies and structures-see Figure 2.

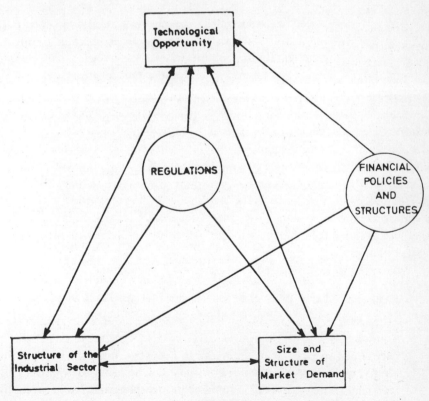

FIGURE 2

Features of a Coherent Policy

As we suggested earlier, there are considerable differences between the innovation policies adopted by different countries. Some countries opt largely for rather general policies designed to create the right environment for innovation. Other countries intervene more directly in the innovation process, promulgating some combination of technologically or industrially non-selective (horizontal) measures and (vertical) measures of technology/industrial selection. More recently, there has been a general trend towards the increasing adoption of policies involving the selection and support of main priority areas. Whatever type or combinations of policies are adopted, it can be stated with some confidence that they should contain at least the following features (the list is by no means exclusive):

Coherence :   The actions of the various institutions involved in policy
                   formulation and implementation should be coordinated
                   in order to avoid the promulgation of contradictory
                   measures, especially between innovation and other poli-
                   cies. More positively, potential synergies must be sought
                   and capitalised upon. Innovation policies and general
                   macro-economic policies must pull together.

Consistency :   Innovatiion policies must be insulated from the dictates
                   of the short-term political cycle. While this might not
                   be too difficult to achieve with tactical policies, it
                   might be more difficult with strategic policies involving
                   major programmes of restructuring. Innovation policy
                   should not be the creature of party dogma.

Flexibility :   While innovation policies should be consistent, they
                   should not at the same time be inflexible. Policies must
                   be capable of responding to changing industrial needs,
                   threats and opportunities. Greater inherent flexibility
                   might be achieved if the policy initiative has in-built
                   "learning by doing". In other words, policy measures
                   should incorporate on-going evaluation, with positive
                   feedback to the policy system in order continuously
                   to improve policy effectiveness.

Complementarity: Policies should not only complement each other, but
                   they should complement also the strategic interest of
                   domestic companies. This means that policymakers

should be aware of the long-term strategic thinking within major national companies.

Realism :   Policy makers must recognise the inherent limitations of public policy and accept them. Over-optimistic expectations, unmet, might result in disillusionment and the termination of promising initiatives. Policies should thus be based on a realistic assessment of industrial potential. Public policy makers should also recognise their own limitations; while, in consultation with industry, public bodies might be involved in the selection of rather broad areas of techno /economic activity, the choice of individual projects is probably best left in the hands of industrial managers.

Finally, it should be emphasised that technological innovation policies, by themselves, are not enough; they must go hand in hand with the appropriate general economic and social policies. Governments must strive to create a favourable overall economic climate (e.g. low interest rates, moderate corporation tax regimes), a favourable social climate (e.g. stimulate the social acceptance of new technology and help overcome social and institutional rigidities and resistances to change), a relatively stable political climate (dramatic political swings create uncertainty) and avoid rapid policy changes (stop-go policies can deter the adoption by firms of the necessary long-term development strategies). In other words, reindustrialisation can only occur effectively when government gets its *whole* "act" together, i.e. achieves overall policy coherence. Well thought out and implemented technological innovation policies are a necessary but not sufficient condition for succesful reindustrialisation.

REFERENCES

AHO, C.M. and ROSEN, F. (1980), "Trends in Technology-Intensive Trade with Special Reference to US Competitiveness", Science and Technology Indicators Conference, OECD, Paris, 15-19 September.

ALLEN, G. (1981), "Industrial Policy and Innovation in Japan" in C. Carter (ed.), *Industrial Policy and Innovation*, Heinemann.

ALLEN, T. et al., (1978), "Government Influence on the Process of Innovation in Europe and Japan", *Research Policy*, vol. 7, pp. 124-49.

FREEMAN, C., CLARK, J. and SOETE, L. (1982), *Unemployment and Technical Innovation*, Frances Pinter.

GIBBONS, M. (1982), "The Evaluation of Government Policies for Innovation", Six Countries Programme Workshop on the Evaluation of Innovation Policies and Instruments, Windsor, UK, 22-23 November.

GOLDBERG, W. (1981), "Explorations Into the Instrumentality of Innovation Policy", Research Report, International Institute of Management, Science Centre, Berlin.

GOLDING, A.M. (1978), "The Influence of Government Procurement on the Development of the Semiconductor Industry in the US and Britain", Six Countries Programme Workshop on Government Procurement Policies and Innovation, Dublin, Eire.

KAPLINSKY, R. (1982), *The Impact of Technical Change on the International Division of Labour : The Case of CAD*, Frances Pinter.

LITTLE, B. (1974), "The Role of Government in Assisting New Product Development", School of Business Administration, University of Western Ontario, London, Canada. Working Paper Series No. 114, March

MENSCH, G. (1979), *Stalemate in Technology*, Cambridge, Mass., Ballinger.

NATIONAL ACADEMY OF SCIENCES, (1980), *Antitrust, Uncertainty and Technological Innovation*, Washington DC.

OTA, (1984), *Commercial Biotechnology : An International Analysis*, Office of Technology Assessment, Congress of the United States, Washington DC.

ROTHWELL, R. (1979), *Government Regulations and Industrial Innovation*, Six Countries Programme on Innovation, c/o Policy Research Group TNO, DELFT, The Netherlands.

——, (1980), "The Impact of Regulation on Innovation : Some US Data", *Technological Forecasting and Social Change*, vol. 17, pp. 7-34.

——, (1981), "Technology, Structural Change and Manufacturing Employment", *OMEGA*, vol. 9, No. 3, pp. 229-245.

——, (1981 b), "Some Indirect Impacts of Regulation on Innovation in the US", *Technological Forecasting and Social Change*, vol. 18, pp. 57-80.

——, (1982), *Evaluating the Effectiveness of Government Innovation Policies*, Six Countries Programme of Innovation, c/o Policy Reserach Group, TNO, DELFT, The Netherlands.

RUBENSTEIN, A. (et. al.), (1977), "Management Perceptions of Government Incentives to Technological Innovation in England, France, West Germany and Japan", *Research Policy*, vol. 6, No. 4.

ROTHWELL R. and ZEGVELD, W. (1981), *Industrial Innovation nand Public Policy*, Frances Pinter.

, (1982), *Innovation and the Small and Medium Sized Firm*, Frances Pinter.

—— , (1985), *Reindustrialisation and Technology*, London, Longman (forthcoming).

SIGURDSON, J. (1984), "Joint Research as a Policy Instrument with Particular Reference to Information Technologies in Japan", SSTC/TIPCE/IFIAS Roundtable Meeting, Guidelines for Science and Technology Policy for Development, Beijing, China, 4 - 8 May, c/o Research Policy Institute, University of Lund, Sweden.

CHAPTER FOUR

TECHNICAL CHANGE AND EMPLOYMENT

By Bruce Williams*

In the third edition of *Principles of Political Economy and Taxation*, published in 1821, Ricardo wrote that he had become "convinced that the substitution of machinery for human labour is often very injurious to the interest of the class of labourers"[1]. In the first two editions he had held that the introduction of labour-saving machinery was in the general interest "accompanied only with that position of incovenience which in most cases attends the removal of capital and labour from one employment to another". The innovating capitalist would make above-normal profits, but as the new machinery came into general use the price of the commodity would fall back to its cost of production. Profits would return to normal, but the capitalist, like the rest of the community, would get the benefits of the innovation in the form of "a greater quantity of comforts and enjoyment".

The substitution of machinery for labour could injure the interests of labourers by redistributing income. Suppose a capital/output ratio of 3 : 1 and a yield on capital of 10 per cent in the initial situation to be changed by a labour-saving invention into a capital/output ratio of 3.5 : 1 and a yield on capital of 12.5 per cent. In the initial situation with an output of 100, labour would receive 70 and capital 30 (10 per cent on 300 capital). In the second situation, if output did not increase to more than 120, labour would receive less than 70. However for outputs beyond 125, labour would receive more than the initial 70, even though it received a smaller share. If, for example, output rose to 140 and capital to 490, labour would receive 79 and capital 61 (12.5 per cent of 490). The expansion of capital and output would affect the yield on capital. If with an output of 200 the yield on capital came back to 10 per cent, labour would receive 130 and capital 70 (10 per cent of 700). That would not be against the interests of labourers, even though labour's share was only 65 per cent compared to 70 per cent in the first situation.

Technical change may or may not bring an increase in the ratio of capital to output. In Britain the capital/output ratio varied within a fairly narrow range between 1855 and 1973. It varied between 4 and 5 which "is hardly larger

---

* The Technical Change Centre, London.
1. Everyman's Library edition, London, 1911, p. 264.

than the margin of error in the capital stock statistics, and if allowance is made for below-capacity working in the inter-war period, the ratio might reasonably be regarded as a long-run constant at a level of just over 4".[2] In the United States, however, the ratio of capital to output fell significantly between 1919 and 1957 from about 4.5 to 2.6. In that period, during which output per man was increased by about 2.5, the average return on capital rose from 6 per cent to 7 per cent, but labour's share rose from 72 per cent to 81 per cent[3]. Such a growth in output depended on substantial technical change and labour-saving innovations which reduced labour per unit output but did not reduce labour's share.

The substitution of machinery for labour could also injure the interests of workers by creating technological unemployment. Ricardo explained his change of mind on the effects of susbtituting machinery for human labour as a consequence of realizing that "the discovery and use of machinery might be followed by a diminution of gross produce" (p. 266). He assumed an initial capital of £ 20,000, of which £ 7,000 was fixed and £ 13,000 circulating. With a profit of 10 per cent on the £ 20,000, the net produce would be £ 2,000, and the gross produce would be £ 15,000 (the £ 2,000 plus the £ 13,000). He then postulated a situation where the capitalist would raise his fixed capital to £ 14,500 — by employing half his workforce to construct a machine worth £ 7,500 — and reduce circulating capital to £ 5,500. Net produce, the 10 per cent on capital employed, would remain at £ 2,000 but gross produce would fall to £ 7,500. That sharp reduction in circulating capital from £ 13,000 to £ 5,500 would result in "distress and poverty".

After deploying this simple arithmetical example to demonstrate a conflict of interests, Ricardo proceeded to undermine the intellectual basis for any opposition to the introduction of machinery. The introduction of labour-saving machinery would reduce commodity prices, add to "the efficiency of the net revenue" — that is, increase the real value of profits — and so lead to an increase in savings. The increase in savings would add to circulating capital and thus increase the demand for labour. Furthermore, he explained, to elucidate the principle, he had supposed a sudden discovery of new machinery which greatly increased fixed relative to circulating capital, "but the truth is that these discoveries are gradual and rather operate in determining the employment of capital which is saved and accumulated than in diverting capital from its actual employment". (p. 270).

2. *British Economic Growth 1856-1973*, by Matthews, Feinstein and Odling-Smee, Oxford, 1982, p. 137.

3. *The Economics of Labour*, by Phelps Brown, Yale University Press, 1962, pp. 225-6.

Marx

The classical economists treated technical change as a force that could counteract an underlying tendency to a falling rate of profit and a stationary state. In Chapter 2 of Book 4 of his *Principles of Political Economy* (1840), J.S. Mill wrote that physical knowledge was advancing more rapidly and in a greater number of directions than in any previous age or generation, and was "converted by practical ingenuity into physical power" more rapidly than at any previous period. But Chapter 4 of that Book 4 was still concerned with "the tendency to a stationary state". That chapter was often referred to during the great depression of the 1930s.

Marx gave technical change a much more dynamic role. In the *Communist Manifesto*, published in the same year as Mill's *Principles*, Marx, with Engels, had written that "the bourgeoisie cannot live without revolutionizing production". In 1857/8 Marx wrote that when available machinery already provided great capabilities, and large industry had already reached a higher stage and all the sciences had been pressed into the service of capital, "the analysis and application of mechanical and chemical laws enabled the machines to perform the same labour as that previously performed by the workers. Invention then becomes a business, and the application of science to direct production becomes a prospect that determines and solicits it"[4].

In *Capital*, Marx expounded a theory of fluctuations in employment in which capital accumulation and technical change are the key factors. In this theory there is normally a reserve army of unemployed which exerts a downward pressure on wages. But "the essential condition for the existence and sway of the bourgeois class is the formation and augmentation of capital", and periodically the rise in the stock of capital, which determines the employment of labour, reduces the reserve army of unemployed. Wages then rise and profits fall, and this checks the accumulation of capital. The fall in profits induces labour-saving inventions and innovations which reduce employment and wages, and so lift profits and capital accumulation. However Marx did not expound a purely cyclical theory of fluctuations. He assumed that technical change would be dominantly labour-saving and that the consequential increase in fixed capital relative to circulating capital would create an underlying tendency to a falling rate of profit and a trend increase in technological unemployment. He assumed also that technical change would bring increasing economies of scale, reduce the strength of competitive pressures to innovation and make the system more unstable.

4. *Grundrisse,* Penguin Books edition, 1973, pp. 703-4.

In his essay *On the Economic Theory of Socialism*, written during the depression of the 1930s, Oskar Lange drew on this part of Marx's theory of capitalist evolution to explain the capitalist crisis, and to underpin his advocacy of a combination of central planning and market mechanisms to restore employment and growth[5]. As in the thirties the current depression has induced a considerable revival of interest in Marx's theory of the relations between technical changes and employment.

Keynes

In "The Economic Possibilities for our Grandchildren", published in 1930, Keynes wrote of an increase in unemployment due to our discovery of the means of economizing the use of labour faster than the discovery of new uses of labour[6]. He predicted that although much more would be heard of technological unemployment in the years to come it would prove to be only a temporary phase of maladjustment. But just how the maladjustment would be cured was not made clear in that essay, though in his *Treatise on Money*, also published in 1930, he wrote that it is easy to understand why fluctuations should occur in the rate of investment in fixed capital and unreservedly accepted Professor Schumpeter's explanation of the major movements[7]. That could be taken to mean that the depression in prices, profits and employment would be reversed by innovations which created new opportunities for profits and attracted a swarm of improvers.

Although Keynes did not bring together his explanations of unemployment, the implication of his "grandchildren" essay is that fluctuations in the extent of technological unemployment are superimposed on a trend increase in the capacity to produce relative to the desire for additional consumption. In the absence of major wars or increases in population, he wrote, periodic revivals of investment and growth would bring the economic problem within sight of solution within a hundred years. The growing gap between rates of growth in the capacity to produce and the desire to consume would be closed by reductions in the supply of labour (as measured by hours of work on offer). On this line of interpretation, technical unemployment would be generated by periodic imbalances between the strength of

5. *Review of Economic Studies* : Part One in vol. IV, No. 1 (October 1936), pp. 53-71; Part Two in vol. IV, No. 2 (February 1937), pp. 123-142.

6. Included in *Essays in Persuasion*, Macmillan, London, 1933.

7. *Treatise on Money*, vol. 2, p. 85, Macmillan, London, 1930. As editor of *The Economic Journal*, Keynes had published Schumpeter's paper, "The Instability of Capitalism", in 1928.

labour-displacing and labour-generating technical change, perhaps complicated by institutionally-determined delays in adjusting hours of work to the underlying tendency for increases in the capacity to produce to grow faster than the demand for additional consumption.

Keynes worried about the social and personal consequences of this secular reduction in the hours of labour demanded. Based on observation of the behaviour of the idle rich, he thought "with dread of the re-adjustment of the habits and the instincts of ordinary men, bred into him for countless generations, which he may be asked to discard in a few decades". In a more optimistic vein he added, in that essay on the economic possibilities for our grandchildren, that with a little more experience we would use the new-found bounty of nature differently, do more for ourselves than the rich had been accustomed to do, and be only too glad to have duties and tasks and routine. "We shall endeavour to spread the bread thin on the butter — to make what work there is still to be done as widely shared as possible. Three-hour shifts or a fifteen-hour week may put off the problem for a great while. For three hours a day is quite enough to satisfy the old Adam in most of us".

The approving reference in the *Treatise on Money* to Schumpeter's explanation of major fluctuations in investment in fixed capital, and the prediction of a growing gap between increases in the capacity to produce and the demand for additional goods, were not superseded by the publication of his *General Theory of Employment, Interest and Money* in 1936. Keynes was concerned to demonstrate how a decline in the inducement to invest could lead to sustained unemployment rather than, as in classical theory, to a redistribution of employment between the capital-goods and consumption-goods sectors of activity via the effects of the rate of interest on decisions to save and invest. That demonstration provided a much better basis for anti-depression measures by Governments than had previously existed, and helped to persuade Governments to adopt full-employment policies after the second world war.

During the sustained post-war boom, much of the credit was given to what came to be known as "Keynesian demand-management policies", though Keynes never implied that demand management would be capable of sustaining employment rates at 97 per cent or more, and did not advocate Government measures to increase money flows to sustain or increase employment except in situations where the increased money flows would soon induce something like matching increases in goods flows.

Automation and Unemployment?

During the discussions of automation in the 1950s there were many writers who formulated Ricardo-type examples of how the introduction of automatic factories could injure the workers. In 1950, Norbert Wiener predicted that within 25 years automation would result in a depression that would make that of the 1930s seem like a pleasant joke. More recently, "the chip" has been heralded as the new technology of wide application that will dwarf the impact of all previous labour-saving technologies, and although annual hours of work have fallen by over 20 per cent since Keynes wrote in 1930, and the average age of entry of the labour market has increased, the view that Government measures to reduce hours further are required to increase employment has gained ground in the European Community.

In *Lloyds Bank Review* of October 1983, Professor James Meade contemplated the *possibility* that Chips and Robots "will make Men less valuable in production relatively to Machinery", and invited his readers to imagine the extreme possibility of robots producing everything, including the manufacture and replacement of each other, with an absolute minimum of labour to attend them. The wages of labour would be transformed into profits earned on robots, and "although output per head might be immensely increased, either competition among workers for the small number of jobs would reduce the real wage absolutely to a very low level or else, if by trade union or similar action the real wage was held up, there would be a large volume of involuntary unemployment among those who were not privileged to get the few jobs available at the fixed rate of pay".

To prevent this neo-Ricardian situation from seriously depressing the income of workers, Meade proposed measures to make each "representative citizen... a representative owner of property as well as a representative potential worker. The national income, including the products of the Robots, would again be more equally divided and the workers' decreased employment would again become voluntary leisure rather than involuntary unemployment". But if the hours of work were less than desired by "representative workers", not all leisure would be voluntary, and chips and robots would have to be held responsible for "disguised" technological unemployment.

Past Effects

Technical change is not a new phenomenon, and an examination of past effects should provide some guidance to a consideration of possible future effects. The distinction between the stone, bronze and iron ages is based on technical change. Museums in China provide ample evidence of a great range of inventions in Ancient China. There was a major technical change in war technology in the fourteenth century when gunpowder was invented. Another invention in the next century, movable type and type metal, had major social consequences. What is new — or relatively new, since the change has been well established now for over 200 years — is the cumulative and self-sustaining nature of technical change.

From the beginning of the Christian era until 1750, world population did not increase by more than five per cent every 100 years. Then, because of changes — which proved to be continuing changes — in the agricultural technologies, population increased between 1750 and 1850 by 60 per cent. Between 1850 and 1950 population growth rose by 130 per cent. Between 1950 and 1980, when population increased by another 80 per cent, the annual growth was even higher. Yet this increase in population was accompanied by increases in output per head and great increases in the number of jobs. Methods of production in old-established industries changed, new industries were established, and the distribution of jobs between agriculture, industry and services was transformed. In the industrialized countries as a whole, employment is now approximately 7 per cent in agriculture, 35 per cent in industry, and 58 per cent in services. In 1850, approximately 50 per cent were in agriculture, 25 per cent in industry and 25 per cent in services.

Since 1870, there has been a great increase in the labour force — by factors of 2 in the UK, 2.5 in Germany, 3 in Japan and 7 in the USA. Such increases have not been accompanied by diminishing marginal returns to effort as followers of Ricardo and Malthus would have expected. Between 1870 and 1979 gross domestic product per head of population grew by a factor of 9 in the UK, 13 in France, 16 in Germany and the USA, and 18 in Japan. GDP per man-hour, as estimated by Maddison, grew by factors of 17 in France and Germany, 29 in Japan, 7 in the UK, and 12 in the USA[8]. In 1870 GDP per man-hour was higher in the UK than in the other countries. In index terms, the productivity growth in Britain was from 100 to 700, in the USA from 88 to 1060, in France and Germany from 54 to 920, in Italy from 55 to 730, and in Japan from 21 to 550.

When real income reaches a certain level, workers choose to take part of the increase in the capacity to produce in the form of shorter hours, so reducing the increase in the demand in goods below the growth in the capa-

8. Maddison, op. cit., Table C10, p. 212.

city to produce at the previous level of hours. As might be expected from its very low level of product per man-hour in 1870 and relatively low level in 1979, the reduction in annual hours worked per person was least in Japan. There the reduction was only 28 per cent. In France and Germany the reduction was 40 per cent. In the USA the reduction was 45 per cent (as it was also in the UK, despite its lower rate of growth and a GDP per man-hour in 1979 at only two-thirds of the level in the US)[9]. It is interesting to note that 50 years after Keynes wrote his essay on "The Economic Possibilities for our Grandchildren" the growth in productivity and the reductions in hours are not inconsistent with his projections for the year 2030.

Although there has been a long-term relationship between increases in output per man and annual hours of work, it has not been a consistent short-term relationship. In the UK, between 1870 and 1980, for example, annual hours fell by one-quarter of the rate of reduction in hours of labour per unit output[10]. But as can be seen in the chart below there were periods when the relationship did not hold at all. There was a sharp reduction after the first world war as the eight-hour day became general, little further reduction during the depressed period between the wars, though product per worker in employment continued to increase, and then an increase in hours after the second world war which reflected an increase in overtime working. Since the mid-fifties the relations between reductions in hours per unit output and annual hours have been closer.

Although in the past hundred years there has been an increase in the proportion of workers affected by fluctuations in rates of growth and employment, there is no clear evidence of a trend increase in the amplitude of fluctuations. Statistics of trade unionists unemployed in Britain from the 1850s indicate that average unemployment was 5 per cent in the fifties and sixties, a little under 4 per cent in the seventies, just over 5.5 per cent in the eighties and just under 4.5 per cent in the nineties. During the fluctuations from 1870 to 1913, peak unemployment was 11.4 per cent in 1879, 10.2 per cent in 1886, 7.5 per cent in 1893 and 7.8 per cent in 1908[11]. There was an increase in unemployment in Britain between the wars. The average level of unemployment rose from less than 5 per cent between 1900 and 1914 to 9.5 per cent in the twenties and over 14 per cent in

9. Ibid., Table C9, p. 211.

10. Philip Armstrong, *Technical Change and Reductions in Life Hours of Work,* The Technical Change Centre, London, 1984.

11. W. Rostow, *British Economy in the Eighteenth Century,* Clarendon Press, Oxford, 1948, pp. 45-48.

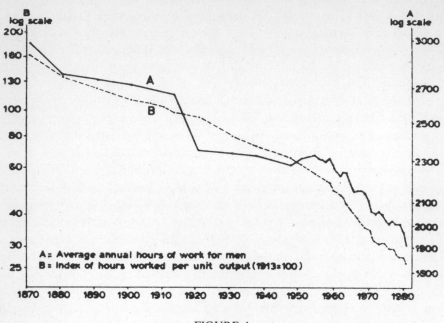

FIGURE 1

Relationship Between Productivity and Average Annual Hours of Work for Men

the thirties. However from 1950 to 1973 unemployment averaged less than 3 per cent, and then 7 per cent from 1974 to 1983.

After 1873 there was a fall in consumer prices in almost all industrial countries. Between 1873 and 1895, consumer prices in Britain and the USA fell by 30 per cent, in Germany by 12 per cent, and in France by 10 per cent. Prices then rose again until the outbreak of the first world war, and then sharply until 1921. Prices then fell until the mid-thirties, rose substantially during the second world war, and continued to rise subsequently. Whereas between 1890 and 1913 the annual average increase in consumer prices was less than 1 per cent, it was over 4 per cent between 1950 and 1973 and almost twice that figure beteen 1973 and 1983.

Growth rates have also fluctuated substantially. In Britain, Belgium, Holland and Switzerland, growth rates between 1870 and 1913 were less than between 1820 and 1870, but in France, Germany, Japan and the USA, growth rates were higher between 1870 and 1913 than in the earlier period. Between 1913 and 1950 the average growth rates in output per head of population in the industrial countries were about 15 per cent less than between 1870 and 1913, though the growth in output per man-hour of those in employment rose by about 12 per cent. Then between 1950 and 1973 the annual growth rate in output per head of population reached more than three times the rates achieved between 1870-1913 and 1913-1950. Between 1973 and 1983 the growth rate fell to less than half the rate achieved between 1950 and 1973, but remained significantly higher than that achieved in earlier periods and in earlier depressions.

The view that it was possible for Governments to maintain full employment was strengthened by the unusually long period of sustained growth in output and employment after the second world war. However that view was undermined by the general decline in growth rates and the rise in unemployment after the mid-seventies. That rise in unemployment is often assumed to be the consequence of the widespread introduction of labour-saving machinery, though the evidence for such an influence would surely be a "productivity upturn" and not, as in fact, a "productivity downturn". The productivity downturn was not caused by any one factor, but declines in the rates of technical change and diffusion were important factors, as they were in the thirties when Keynes wrote of an increase in unemployment due to a lag in the discovery of new uses for labour. However, the current depression differs substantially from that of the thirties in at least two respects. The first difference, due presumably to the much larger proportion of the workforce engaged in R & D, is the higher level of productivity increases. The second difference, due in considerable measure to confidence in the power and will of Governments to accommodate price levels to wage levels in the interests of full employment, is a continued rise instead of, as previously, a fall in prices.

Long Cycles?

The recurrence of severe unemployment about 50 years after the great depression revived interest in the literature on long cycles. In 1913 van Gelderen and Pareto drew attention to the signs of something like 50-year cycles in the movement of prices and interest rates. Then in 1926 Kondratieff, director of the Business Cycle Research Institute in Moscow, argued from the movements since 1779 in wholesale prices, wages, interest rates, bank deposits, the values of foreign trade, coal production in England, coal consumption in France, and pig iron and lead production in England, that there were signs of long cycles of 40-60 years, medium cycles of 7-10 years, and short cycles of 3-4 years[12]. It is not possible to draw any signifi-

12. N.D. Kondratieff, "Die langer Wellen der Konjunktur", in *Archiv für Sozial-*

cant condusions about fluctuations in output and employment from the physical indicators used by Kondratieff. His main evidence for the existence of long waves came from the interrelated monetary indicators. It was not until the twenties in Britain, and in the thirties more generally, that there was a substantial rise in unemployment in the downswing of a period which Kondratieff and followers identified as the downswings of long waves.

In 1930, Kuznets published the results of his more systematic analysis of a larger number of price and output series, and concluded that there had been secular variations in production similar in most cases to those in prices, that the complete swing was only 22 years for production and 23 for prices, but that the evidence would not support a conclusion that these variations were cycles[13].

Despite Kuznets' evidence, in 1939 Schumpeter wrote of 50-year long cycles, with four stages of prosperity, recession, depression and revival, as if the existence of regular long cycles had been proven[14]. The main significance of Schumpeter's work on cycles was the extension of his earlier work on the role of entrepreneurs in economic development, and the identification of the role of particular new technologies[15]. His starting point was the tendency of competition to eliminate profits and the possibility of re-creating profits by innovations. The profits of successful investments in innovation would encourage imitations and improvers and substantial further investment activity. The profits of innovation would then be competed away and the investment boom would tail off. During the subsequent depression the least efficient firms and equipment would be eliminated, the new processes and products would be improved and consolidated, and the economic system would begin to settle down at a higher level of output per head — until disturbed by new entrepreneurial activity which would generate a new long wave.

In his first long cycle from about 1770 to 1827, the upswing and prosperity phases, 1770-1800, were based on technical innovations in metallurgy, textile machinery and steam power. In his second long cycle, 1828 to 1885, the upswing and prosperity were based on improvements in metallurgy and machinery which made possible the use of steam engines in tran-

wissenschaft und Sozialpolitik, 1926, pp.573-609 (English translation by Stolper, Review of Economic Statistics, November 1935).

13. S. Kuznets, *Secular Movements iu Prices and Froduction*, Houghton Mifflin, Boston, 1930.

14. J.A. Schumpeter, *Business Cycles*, McGraw Hill, New York, 1939.

15. J.A. Schumpeter, *Theorie der Wirtschaftlichen Entwicklung*, Duncker und Humboldt, Leipzig, 1912. (English translation, Harvard, 1934).

sport. The third cycle, 1886 to 1937, was based on the rise of the science-based electrical industry, the transformation of the chemical industry by organized research, and the internal combustion engine. The generation and the strength of the fourth long wave, dated by followers of Schumpeter as 1938 to ?1989, was based on innovations in synthetics, drugs, jet aircraft, electronics, pesticides and animal and plant breeding. Some followers of Kondratieff and Schumpeter confidently predict a fifth long wave based on widespread applications of information technology and biotechnology.

It seems likely that the influence of Schumpeter's *Business Cycles* on economists who were later involved in the formulation of plans for postwar reconstruction and in the conduct of economic policy after the war, was reduced by its publication just before the second world war, and at a time when macro-economists were still preoccupied with the analysis and policy prescriptions in Keynes' General Theory. Two other factors contributed to its lack of influence until the 1970s. The first factor was the absence either of clear statistical evidence for the existence of regular short, medium and long cycles, or of a plausible theory of just why innovations of many different kinds — some deriving from exogenous and some from endogenous inventions, some presenting few problems of application outside the industry and some dependent on complementary innovations elsewhere — had generated and would continue to generate fairly regular cycles. The second factor was the development of the thesis in his *Capitalism, Socialism and Democracy* (1943) that with the growth of organized industrial research and development, large corporations would be increasingly able to make of innovation a routine function.

Gerhard Mensch did not accept Schumpeter's view that large corporations with large R & D departments had become capable of producing innovations to order and were turning innovation into a routine function. In *Stalemate in Technology: Innovations Overcome the Depression*[16] — the main thesis of which was conceived and developed in 1970-71 when, he wrote in the preface, economists still projected incessant economic growth, and science and technology policy were expected to produce a steady flow of opportunities for innovation — he presented a metamorphosis model of socio-economic change. This model was based on the propositions that basic innovations which establish new branches of industry and radical improvements which rejuvenate existing branches come in rushes, and that as the technological potentialities of the major innovations become fully used,

16. The original German edition was published in 1975. The English language edition was published by Ballinger at Cambridge, Mass., in 1979.

due to "power, inertia and other decisive reasons", large companies no longer want to engage in innovation competition with the same intensity as before. The preference for pseudo-innovations over basic innovations leads to stalemate (stagnation). Then "under pressure from high unemployment and under-utilized capital, opposition to and reservations about untried, risky new ideas disappear", and a surge of basic innovations ends the stagnation[17].

He concludes from his reading of the pattern of past delays in "the transfer of knowledge followed by acceleration because of the pressure from the dammed-up demand for technological basic innovations" that a new surge will begin in earnest after 1984 and that "approximately two-thirds of the technological basic innovations that will be produced in the second half of the twentieth century will occur in the decade around 1989" (p. 197).

To test his basic thesis Mensch relied mainly on a selection from the list of inventions in the first edition of *The Sources of Invention* by Jewkes, Sawers and Stillerman[18]. Freeman, Clark and Soete have rightly criticized Mensch for reliance on the list in the first edition which did not do justice to the inventions and innovations in the 1950s and 60s, and for his exclusion of those inventions for which the time lag between invention and innovation is less than ten years. Use of the full list, particularly in the second edition of *The Sources of Invention*, considerably weakens the evidence for "bunching" during the depression of the 1930s.

Freeman, Clark and Soete showed that there was a substantial number of major innovations in the fifties and sixties, and concluded that "the significance of "swarming" or "bunching" relates more to the diffusion process and to combinations of sets of basic innovations with many improvement innovations rather than the initial appearance of the individual basic innovations themselves"[19]. In any case Mensch tried to prove too much. His theory that about two-thirds of the major innovations that generate new long waves are concentrated in a decade during depression is inherently implausible and is not needed to explain recovery from stagnation. All that is required is sufficient innovation to lay the foundations for confidence in a recovery process which would then encourage further investment in innovation.

To explain discontinuities in the innovation process, Freeman and his associates placed much greater emphasis on "bunching" during the diffusion

17. Op. cit., p. 178.
18. Jewkes, Sawers, and Stillerman, *The Sources of Invention*, London, Macmillan, 1958, revised edition 1969.
19. Freeman, Clark and Soete, *Unemployment and Technical Innovation*, London, Frances Pinter, 1982, p. 45.

process, and then identified three other significant factors : first, the tendency for any pattern of innovation to shift over time from relatively labour-intensive product innovations to labour-saving process innovations; second, the tendency of the rate of profit to fall due to the growth of capacity, the saturation of markets, and the growth of capital intensity; and third, the increase in the bargaining strength of labour following long periods of full employment. But their account of the band-wagon effect and the recession and depression phases is much more satisfactory than their account of the generation of new waves. Indeed their account of conditions that encourage or discourage innovations implies that recovery requires exogenous stimuli. They expect the depth of depression to inhibit or delay basic inventions, and "because of the diminishing flexibility of market economies generally, and the important role of the public sector in R & D and innovations policies, the role of public policy is crucial. But some exogenous stimulus to the system, which helps to restore the general level of profitability and improve expectations generally, may also be necessary" (p. 81)[20].

At the Massachusetts Institute of Technology, J.W. Forrester has developed a Systems Dynamic National Model "to represent the physical and human process found in any national economy". Forrester believes that this model explains the reasons for the long waves ending in "the severe depressions of the 1830s, the 1890s, the 1930s, the and 1980s. The process involves an over-building of the capital sectors beyond the capital output rate needed for long-term equilibrium. In the process capital plant throughout the economy is over-built beyond the level justified by the marginal productivity of capital. This over-expansion leads to a depression during which the excess capital plant is physically and financially depreciated until the stage has been cleared for a new era of rebuilding[21].

This model generates long waves without any technological change, and Forrester sees the long wave as compressing technological change into certain time intervals and as altering the opportunities for innovation : for the last half of a 30-year upswing of a long wave, "radical innovation remains outside the circle of acceptance". In this respect Forrester's explanation has much in common with Mensch. However it is very doubtful whether the Forrester

20. In his "Background Material for a Meeting on Long Waves, Depression and Innovation", IIASA, August 1983, Freeman referred to the need for a new international framework and added the comment that the development of such a "framework for expansion may prove to be the most difficult problem confronting the world economy if there is to be a fifth Kondratieff upswing".

21. A convenient reference is J.W. Forrester, "Innovation and Economic Change" pp. 126-134 in *Long Waves in the World Economy*, ed. C.Freeman, London, Frances Pinter, 1984.

mechanism could have been sustained for over 150 years without technical change to disrupt the damping mechanism that market processes do provide. Another problem with the Forrester explanation arises from the importance of capital formation data which are notoriously difficult to assemble and interpret. For Britain, Coombs has shown that the Forrester explanation does not fit relative movements in the capital and consumption sectors of the British economy in the last 60 years[22].

Among those who support the view that there are long waves, there are considerable differences of opinion on the dating and the role of technical change. In *The World Economy, History and Prospect*, Rostow emphasizes the rates of change in the profitability of producing foodstuffs, raw materials and commodities. As a consequence of this emphasis, Rostow's dating of upswing and downswing differs, and at times differs substantially, from writers who concentrate on innovations in the secondary and service sectors. He treats 1951-73 as the downswing of a fourth Kondratieff[23].

How Much Technological Unemployment?

Technical change has led to substantial increases in gross national product per head and considerable reductions in hours of work per year and in the number of years spent in the labour force. There have been substantial costs attached to these benefits, but the benefits and costs have not been shared evenly between people or between regions or between years. Currently there are fears that the costs of technical change are rising relative to the benefits, and in particular that technical change will generate more serious employment problems than in the past.

Technical change consists of process innovations and improvements and product innovations and improvements. Process innovations may save labour, or working capital or fixed capital per unit output, or some combination of these factors. Product innovations may displace in whole or in part older products, or open up new markets.

In agriculture, technical change has been responsible for a very large displacement of labour. There have been some product innovations, but process innovations which raise crop and animal yields per acre have been dominant. Improvements in seeds, breeds, fertilizers and pesticides have greatly increased output per unit labour input, and innovations in farm ma-

22. In Freeman, op. cit., pp. 115-125.

23. W. Rostow, *The World Economy, History and Prospect*, London, Macmillan, 1978.

chinery have led to widespread substitution of capital for labour. Reductions in labour per unit output by x per cent lead to reductions in the employment of workers whenever the sales of the product increase by less than x per cent. In agriculture, the percentage expansion in sales has been very much less than the percentage reduction in labour per unit output. There has been little product innovation, and in any case the capacity of the human stomach is limited.

The effects of technical change on employment in industry have been very different, mainly because of the great range of product innovations. There have of course been many industrial products where employment has fallen. There is now a considerable literature on product life cycles. Many new products when first introduced are labour-intensive but are made much less so by a succession of process innovations. Motor cars, for example, when first introduced were very expensive and labour-intensive, but process innovations − at first standardization and the assembly line and later a series of moves to automation − brought great reductions in labour per unit output and price. For a long time the percentage increases in output that followed reductions in price exceeded the percentage reductions in labour per unit output, and process innovations as well as the original and later product innovations increased the demand for labour. Recently however the degree of market saturation has been such that percentage increases in sales have been less than the percentage reductions in labour per unit output. Future changes in the trend demand for labour in industry will depend on the extent to which the demand for industrial products is sustained by product innovations and on the rate and nature of process innovations (whether predominantly labour-saving or capital-saving).

In the service sector the rise in personal post-tax incomes, and in Government revenues following increases in the taxable capacity of persons and companies, has led to a great expansion in employment and in the proportion of employment in services. A significant part of this expansion has been due to increases in Government expenditure on education and health services.

There are fears that further growth in employment in the services will be prevented by the labour-saving potentialities of innovations in information technology and by the decisions of Governments to prevent further rises in, and perhaps to reduce, the proportion on the national income absorbed in public expenditure. It is by no means certain that information technology will prove to have a general labour-saving effect or that Governments will prevent further increases in public expenditure relative to GNP. But if information technology has a net labour-saving effect the hours of work consistent with high employment will be reduced. And if Governments do reduce their demands for labour in services, the hours of work consistent with full employment will be further reduced unless the lower burden of tax brings an offsetting increase in the private demand for goods and services.

We do not know what the rate of technical change in the future will be — whether a stage of diminishing marginal returns to further R & D has been reached, or whether the increase in the scale of R & D and the improvement in scientific instruments, computers and programmes will have the opposite effect. We do not know whether technical change will be tilted further towards net labour-saving new technologies, a possibility which Meade contemplated in his "Chips and Robots" essay. We do not know whether there will be a change in the preferred rates of trade-off between more leisure and more income.

When dealing with possible future events it is both natural and sensible to think in terms of current forces which may cause departures from known events. There are however considerable differences between writers in the time-scale of their thinking. Thus, on the basis of his appraisal of the work of Kondratieff and Schumpeter on "long cycles", Mensch predicted a substantial depression and then a recovery from the mid-1980s. However economists and economic historians are not agreed on the dating or even on the existence of long cycles or waves, and as there is not as yet a satisfactory explanation of the bunching of major innovations, it would be unwise to base policy on his prediction. Mensch himself states that his "bold projection into the future" is conditional on "the assumption that rhythmical pattern in the interplay of stagnation and innovation will continue to evolve in the next 20 years as it has done in the last 300 years or so", though he displays a considerable degree of confidence in his bold projection.

Other writers have placed more emphasis on recent changes in technology — in particular on developments in computers and microprocessors — which they predict will bring a major and lasting shift in the nature of technical change. If that does happen, the problem of adjusting skills, hours of work and the location of work to keep unemployment down to tolerable levels, will be greater than in the past. However the basis for believing that there has been such a change is not very strong, and, as noted above, an acceleration in the rate of increase in output per employed worker that would follow such a major tilt to technical change has not appeared.

In the past, the periods when the labour-creating influences in technical change were weak proved to be relatively short. Market forces did operate to check this fall in labour-creating influences relative to labour-saving influences, and, even though our knowledge of the nature and speed of operation of these forces is incomplete, there is reason to expect their continuance. In their explanation of the downturns, Freeman, Clark and Soete argue that as new industries and technologies mature, and profit margins contract with the growth of capacity and the increase in the bargaining strength of employees, firms have an incentive to concentrate on cost-saving innovations, and that in consequence the growth of employment slows down or stops altogether. But as yields from R & D which is concentrated on cost-saving process innovations in existing lines decline, some firms at least will decide to redirect their R & D towards new products, or to major process innovations in fields where demand is likely to be responsive to price reductions, with a greater potential for higher profits. The other corrective factor, which is also likely to operate in the future as in the past, is the rise in the inducement to invest that follows reductions in excess capacity and cost-push inflation during the course of the depression.

The policy implications of this analysis are easy to state and very difficult to put into effect.

Measures to reduce the high level of inflation that followed the productivity downturn of the early seventies have intensified the current depression, and it is therefore important to establish incomes policies which prevent increases in wages and salaries from getting considerably out of line with productivity increases.

Although there has not been a trend increase in technological unemployment, fluctuations in innovations based on new or improved technologies have created periods of serious unemployment. The OECD report of 1980 on *Technical Change and Economic Policy* recommended as a stabilizing measure that Governments should give more support to strategic and exploratory development work on enabling technologies. Such support should increase, particularly for small and new firms, technology options – which at the moment are biased by the dominance of large firms in the conduct of industrial R & D – and reduce the time lags and risks involved in further innovation. The choice of the enabling technologies and the nature of the support present difficult problems, which at the moment are being tested in the field of information technology in the ESPRIT programme in Europe and the Alvey programme in the United Kingdom.

The creation and effective use of technical opportunities for innovation depend in considerable measure on acquired mental and manual skills. Speed in identifying the skills that new technologies call for, and the prompt provision of appropriate training arrangements, would also reduce the risks of innovation when in the interests of counter-cyclical activities it is most important to reduce such risks.

Measures to combine the provision of risk capital with management services for new and small firms, particularly at times when innovations that will sustain employment are showing signs of slowing down, should also be promoted to reduce the depth and length of periods of stagnation.

REFERENCES

ARMSTRONG, PH. (1984), *Technical Change and Reductions in Life Hours of Work*, The Technical Change Centre, London.

BROWN, E.H.P. (1962), *The Economics of Labour*, Yale University Press.

COOMBS, R.W. (1984), "Innovation, Automation and the Long-wave Theory", in Christopher Freeman (ed.), *Long Waves in the World Economy*, Frances Pinter, pp. 115-125.

FORRESTER, J.W. (1984), "Innovation and Economic Change", in Christopher Freeman (ed.), *op. cit.*, pp. 126-134.

FREEMAN, C., CLARK, J.A. and SOÈTE, L.L.G. (1982), *Unemployment and Technical Innovation*, Frances Pinter.

JEWKES, J., SAWERS, D. and STILLERMAN, R. (1958, revised 1969). *The Sources of Invention*, Macmillan.

KEYNES, J.M. (1930), *Treatise on Money*, Macmillan.

—— (1933), "The Economic Possibilities for our Grandchildren", in *Essays in Persuasion*, Macmillan, p. 142.

—— (1936), *General Theory of Employment, Interest and Money*, Macmillan.

KONDRATIEFF, N.D. (1926), "Die langer Wellen der Konjunktur", in *Archiv fur Sozialwissenschaft und Sozialpolitick*, vol. 56, pp. 573-609. (English translation by W.F. Stopler, "Long Waves in Economic Life", *Review of Economic Statistics*, vol. 17, No. 6, November 1935, pp. 105-115).

KUZNETS, S. (1930), *Secular Movements in Prices and Production*, Houghton Mifflin, Boston.

LANGE, O. (1936 & 1937), "On the Economic Theory of Socialism", *Review of Economic Studies*, Part One in vol. IV, No. 1 (October 1936), pp. 53-71; Part Two in vol. IV, No 2 (February 1937), pp. 123-142.

MADDISON, A. (1982), *Phases of Capitalistic Development*, Oxford University Press.

MATTHEWS, R.C.O., FEINSTEIN, C.H. and ODLING-SMEE, J.C. (1982), *British Economic Growth 1856-1973,* Oxford University Press.

MARX, K. (1973), *Grundrisse*, Penguin Books edition, London.

MARX, K. and ENGELS, F., (1967), *The Communist Manifesto*, Penguin Books edition, London.

MEADE, J. (1983), "A New Keynesian Approach to Full Employment", in *Lloyds Bank Review*, Number 150, pp. 1-18.

MENSCH, G. (1979), *Stalemate in Technology: Innovations Overcome the Depression*, Ballinger, Cambridge, Mass. (Original German edition, 1975).

MILL, J.S. (1865), *Principles of Political Economy*, People's Edition, London.

ORGANISATION FOR ECONOMIC CO-OPERATION and DEVELOPMENT (1980), *Technical Change and Economic Policy*, Paris.

RICARDO, D. (1911), *Principles of Political Economy and Taxation*, Everyman's Library Edition, London.

ROSTOW, W.W. (1948), *British Economy in the Eighteenth Century*, Clarendon Press, Oxford.

—— (1978), *The World Economy, History and Prospect*, Macmillan.

SCHUMPETER, J.A. (1912), *Theorie der Wirtschaftlichen Entwicklung*, Duncker und Humbold, Leipzig. (English translation, *Theory of Economic Development*, Harvard University Press, 1934).

—— (1939), *Business Cycles*, McGraw Hill, New York.

—— (1943), *Capitalism, Socialism and Democracy*, Harper and Row, New York.

CHAPTER FIVE

THE TECHNOLOGICAL BALANCE OF PAYMENTS IN PERSPECTIVE

By Michèle Ledić* and Aubrey Silberston**

Introduction

The technological balance of payments comprises flows of payments for technological transfers between countries, usually payments made and received by individual firms. The figures are not large, by the standards of other balance of payments flows, but it can confidently be said that the importance of technology transfers between countries is much greater than the size of the figures suggests. The figures are however of considerable interest in themselves, and this essay attempts to set them out and to analyse them.

The technological balance of payments comprises receipts from overseas royalties and similar payments, and corresponding expenditure overseas. Royalties can be broadly divided into technological and artistic, and we shall be concerned here with technological royalties only. In the UK case, these include payments arising from licences, patents, trademarks, designs, copyrights, manufacturing rights, the use of technical "know-how" and technical assistance. Other countries include somewhat different combinations of items, including management fees, as in the case of the USA.

These are a number of different ways of transferring technology, and the technological balance of payments reflects only part of the process. It does not, for example, take account of technical progress embodied in the export and import of products. Nor does it include assistance given by a firm to an affiliated firm overseas unless royalties or fees are paid for this. An important part of the international transfer of technology takes place through direct investment, and it has been argued (Pavitt 1981) that in quantitative terms the value transferred through this channel exceeds that transferred through patents and licences.

The technological balance of payments therefore covers only a part of the international transfer of technology. Moreover, technological balance

* University of Zagreb, Faculty of Economics.
** Imperial College of Science and Technology, London.

This essay is based on an earlier paper (Ledić and Silberston 1983). We have made extensive revisions and have added a new section.

of payments statistics are deficient in many countries and there is a considerable problem of consistency between countries. The figures that are available, however, represent significant flows of technology and are therefore of much interest. During the last few years, more and more relevant figures have become available (many of them through the efforts of the OECD) and we are now able to study trends and to make comparisons in a way that was not possible until recently.

The purpose of this essay is partly to analyse trends, particularly in the period since 1970, on a world-wide basis. Attention will be paid both to trade between developed countries and to that between developed and developing countries. A primary focus of interest will be the relationship between affiliated firms, i.e. between parent firms and their subsidiaries. It will be suggested that royalty payments by subsidiaries to parents are used as a partial substitute for the transfer of profits, and are at a higher level than is found between independent firms. We ask how far it is possible to estimate the "right" level of such royalties, and we try to place in perspective the order of magnitude of such flows.

How Royalties Arise

One of the main ways in which royalties are incurred is as a result of the granting of licences on patents (Taylor and Silberston 1973). Patents have a limited life. In the UK, for example, until the 1977 Patents Act their life was 16 years from the date of filing. The period was lengthened to twenty years in 1977, in line with EEC practice. The effective life of a licence is however normally a good deal shorter than twenty years. It may take several years for a patent to be granted, and a further period may elapse before its value has been demonstrated sufficiently for overseas firms to wish to take out licences. Where an overseas firm is associated with the patent holder, the licence may be granted early in the life of the patent, but even so it is unlikely that the typical licence will have more than fifteen years to run.

"Know-how" agreements are often associated with licence agreements. Although patent specifications are meant to give sufficient information for anyone skilled in the relevant art to understand what is involved, this is often not the case in practice, especially with process patents. When a licence is granted, an agreement is often made at the same time to supply the "know-how" needed to operate the patent effectively. Know-how agreements are mainly associated with licence agreements, although they are sometimes made independently. They usually lapse when the licence agreement lapses, and often do so before this.

It follows from what has been said that the receipts and payments included in a country's technological balance of payments are likely to reflect the historical situation relating to technology transfer. Agreements in force could have been made as long as twenty years ago. The average payment or receipt may perhaps result from an agreement made seven or eight years earlier. The time lags involved are therefore likely to be sizeable.

Royalties are usually related to the volume of sales of patented products, or of products made from patented processes. Sometimes an initial lump-sum, or a lump-sum per annum, is also paid — especially when there is political instability or doubt about the ability of the licensee to produce output reliably — but a royalty on sales usually accounts for the bulk of the payments made. Royalties are normally calculated as a percentage of the selling price. The percentages paid vary widely, although there are differences between industries which seem to be well-established.

In so far as royalty payments are based on the value of output of a patented product, or of a product made by a patented process they will (at constant prices) fluctuate with the current volume of output of the product concerned, even though based on agreements made several years before. We try to examine the relationship between the level of royalties and the level of output later in this essay.

A particular problem concerns payments of royalties between subsidiary companies and their parents. These are important quantitatively, as will be seen. The use of licensing payments in connection with technology transfer within a group of companies may however be primarily a response to the tax situation in the relevant countries. Such payments may bear little relation to the output produced under licence, and may be calculated on a rather arbitrary basis, for example as a percentage of subsidiaries' total sales (Taylor and Silberston 1973). As will be seen, the figures suggest that something like this may often take place.

Where royalties are paid as the result of patent licences, the patents themselves must have been granted in the countries concerned. In the absence of patents, know-how agreements would have to be made, or arrangements for payments for technical assistance. A country with an inadequate or inefficient patent system is therefore likely to have to make a high proportion of its payments for new technology in forms other than those of royalty payments. Developing countries may well be in this position. It will be seen later that payments for technical assistance by developing countries are relatively more important than by developed countries, and royalty payments relatively less important. This may be as much a reflection of weak patent systems in developing countries as of their particular needs.

We now turn to an examination of the technological balance of payments position of different groups of countries.

The Surplus Countries

Only three countries have a persistent surplus of any magnitude in their technological balance of payments – the United States, Switzerland and the United Kingdom.

Tables 1 to 3 set out the figures in constant prices, for the main countries from 1970 to the latest available year. The United States is by far the dominant country, followed by Switzerland and the United Kingdom. Receipts for the USA during the period were more than double those for Switzerland and up to ten times those for the UK. Expenditure for the USA was no higher than that for the UK and below that for Switzerland. The result was to give the USA a massive net balance, rising to almost $ 5,000 million at 1975 prices, compared with up to $ 1,500 million for Switzerland and $ 130 million or less for the UK.

Breakdowns are available for the United States and the UK showing receipts and payments by affiliated and unaffiliated firms. These are set out in Tables 5 and 6. In the case of the USA, some 80% of the receipts of royalties, licence and management fees come from affiliated firms, while about half of payments goes to these firms. In the case of the UK, on the other hand, receipts from affiliated firms are below receipts from unaffiliated firms, but payments going to affiliated firms are much greater than those going to unaffiliated firms. The USA has a surplus with both categories, while the UK has a deficit with affiliated firms, offset by its surplus with unaffiliated firms.

These contrasted results are to a considerable extent mirror images. If UK receipts and expenditure are analysed by broad regions (Table 5), it appears that in 1980 the UK had a deficit of some £ 127 million with North America (largely the USA). This consisted of a deficit with affiliated firms of £ 144 million and a surplus with unaffiliated firms of £ 17 million. With the EEC and EFTA, on the other hand, the UK had a surplus of £ 26 million : affiliated firms accounted for the larger part of receipts and payments, but the UK's surplus with these was much less than with unaffiliated firms. With the rest of the world, the UK had a surplus of £ 150 million, split approximately equally between affiliated and unaffiliated firms. North America accounted for one-third of total UK receipts, but for three-quarters of payments : most payments were to affiliated firms.

These figures for the UK demonstrate the strong technological links between the UK and the USA. Subsidiaries of US multinationals in the UK clearly draw heavily on their parent companies for technical expertise. Licence payments are to be expected in cases where the UK subsidiary is not fully owned by its US parent, but it seems obvious from the figures that there must be substantial licence payments by fully-owned subsidiaries, since many of the largest subsidiaries are of this type. In these cases it would be possible for technical knowledge to be shared without explicit recourse to licence payments. The fact that such payments are made is clearly the result of policy decisions by the firms concerned, perhaps connected with tax considerations. Public relations reasons may also play a part.

Looking at it from the point of view of the USA, the great majority of transactions take place between United States firms and their subsidiaries overseas — some 80% of receipts and two-thirds of payments (Bond 1981). The majority of US transactions (almost 70%) occur in manufacturing industries, and within manufacturing the machinery and chemical industries are responsible for one-half to two-thirds of all receipts (the same is true for the UK). United States payments for technical knowledge consist of a small number of relatively large transactions, while receipts typically consist of a large number of transactions (Bond 1981).

US technology is transferred primarily to developed countries (responsible for 80% of US receipts), with Canada, the UK and Japan accounting for almost half of receipts in recent years. Japan has been increasing relatively in importance, while Europe has been decreasing, as have developing countries. The proportion of US know-how being sent to developing countries was 25% in 1968. It fell to 20% in 1978, although the absolute amount increased.

Most US transfers to Canada (over 90%) and to developing countries (85%) are through United States subsidiaries. Most US-Japanese transfers, on the other hand, have historically been between independent firms. In recent years, however, Japanese policy towards foreign capital investment has become more liberal, and transactions between affiliated firms have increased from less than 30% of US receipts from Japan in 1967 to over half in 1978.

In general, trends in US royalty and fee data are paralleled by trends in US direct investment abroad, both as regards countries and types of industry. International trade data provide additional support for the conclusions drawn from the royalty and fee data. They show that the US enjoys an overall favourable balance of trade in R & D-intensive manufactured products (Bond 1981).

We do not have detailed figures for the technological balance of payments of Switzerland, although we have seen that nearly all the payments from the UK to Switzerland are to affiliated firms. It seems clear that much of Switzerland's surplus arises from the large international Swiss companies which are based there, notably in pharmaceuticals and electrical engineering. Hoff-man-LaRoche, the pharmaceutical firm, is a prominent example. In addition, Switzerland is the location of the European head office of a number of US multinationals. Many of these appear to receive royalty payments in Switzer-land, no doubt as a matter of policy, and possibly because of tax considera-tions.

The Deficit Countries - 1) Developed Countries

All the leading developed countries spend a substantial amount on techno-logical royalties, including the UK, Switzerland and the USA. Indeed, the magnitudes are similar. In 1978, for example, expenditure by France, West Germany, the Netherlands, Italy and Japan varied between $ 530 million (at 1975 prices) and $ 870 million, while expenditure by the UK was $ 500 million and by the USA $ 511 million. The difference between the two main surplus contries, the USA and Switzerland, and the other developed countries was in the level of their receipts. These varied in 1978, for example, between not much more than $ 100 million (at 1975 prices) in the case of Italy, to over $ 600 million in the case of France, while the UK had receipts of $ 590 million. The comparable US figure was nearly $ 5,000 million, and the Swiss figure $ 2,000 million.

The effect of these differences has been to leave many of the largest deve-loped countries with deficits in their technological balance of payments. There have, however, been some interesting developments during recent years. Receipts have grown markedly in real terms in all the leading countries, with the exception of West Germany. The most dramatic increases have been in Japan, France and Italy — the first more than doubling its receipts during the 1970s and the last two doubling them. Expenditure by France rose substantially until the mid-1970s, but Japanese expenditure fell sharply after 1972. The result has been a narrowing of the gap for Japan and France, while the West German and Italian deficits have stayed at much the same level. On present trends, France and Japan may show a surplus within a few years.

The fact that both Japan and France are still in deficit, as is West Ger-many also, shows that relatively successful industrial progress is perfectly compatible with a deficit in the technological balance of payments. The UK, which still has a surplus, has done comparatively less well in terms of indu-strial growth, and the same has been true of the USA in recent years. Japan, in particular, has obviously made excellent use of foreign developments in technology, especially those arising in the USA. As will be seen, it is unlikely that Japan could have done this without a great deal of its own expenditure on R & D activities.

How far the technological balance of payments of developed countries reflects the true level of technology transfer between them is a question that has been asked more than once. There is in particular some suspicion that royalty payments between the subsidiaries of US multinationals and their parent companies are higher than they should be, and are used as a convenient substitute for transmitting profits internationally. In the case of Germany, for example, more than three-quarters of total royalty payments abroad in 1979 went to affiliated firms, while these firms contributed less than 7% to receipts (Table 7). At least one German author has been led to the conclusion that "the figures of the patents and licences account prove to be far beyond any reasonable proportions" (Horn 1981). Similar suspicions have been raised in the case of Italy (Scarda and Sirilli 1981). We return to this subject later.

The Deficit Countries – 2) Developing Countries

It is well known that the bulk of research and development is carried out in developed countries. The developed countries are also the home of the multinational corporations. It is therefore to be expected that the great majority of patents should originate in developed countries (Schiffel and Kitti 1978), and that these countries should be the main source of technological and management expertise. Most developing countries of any size have adopted a strategy of industrialisation during the last twenty to thirty years. In the course of industrialisation they have tended to adopt Western technology, especially when multinational firms have set up subsidiaries there. The developing countries have therefore been the recipient of technology arising in the developed countries, and figures for their technological balance of payments reflect this.

Comprehensive figures for the technological balance of payments of developing countries are not readily available, but data from several of the leading developed countries make the position clear. The figures for Germany, for example, are given in Table 7. Developing countries accounted for under 1% of total German expenditure on patents etc. during the 1970s, but receipts from them accounted for between 18% and 30% (with the relative importance declining during the period). Affiliated firms were of comparatively little importance.

It has already been pointed out that 20 - 25% of US receipts from licences and fees came from developing countries during the period 1968 - 78, and that 85% of these payments were from US affiliated companies. US expenditure on licences etc. from developing countries is very small (Table 6), although in recent years it has risen.

The experience of the UK corroborates that of other developed countries. The figures for 1980 are given in Table 5. Receipts from developing countries amounted to 19% of the total ,but expenditure for 1% only. Receipts from unaffiliated firms accounted for some 60% of the total from developing countries, and payments went largely to unaffiliated firms also. In this respect the pattern for both receipts and payments was different from that for any of the other groups of countries shown in Table 5, since in other groups the picture was dominated by affiliated firms.

The type of technology transaction appears to depend on the extent of differences in the technical level of buyers and sellers. As between OECD countries at similar technical levels, for example, trade seems to concentrate on patents and licences and related technical services, especially at an early stage of a new product or process. With developing countries, payments for industrial property rights seem to be comparatively less important than continuing technical assistance. The differences are illustrated in Table 8 — which gives figures for France, Italy and the United States. In all three countries receipts for patents and licences come largely from developed countries, but receipts for technical assistance are proportionately more important than patents or licences in developing countries.

Further figures from Italy reinforce this impression. Italian experience suggests that, especially in the case of developing countries, it is often true that, along with the transfer of plant, a number of other services are supplied by the selling firm : for example, the supply of engineering services, technical assistance, a training period for personnel, and possibly commercial assistance also. In 1979 Italian firms earned 260 billion lire from selling plants abroad (including associates services), as compared with receipts from licences etc. of 145 billion lire. The bulk of receipts for plants came from developing and newly industrialising countries. The data for plant sales show a high positive correlation with receipts from licences etc. from the same category of countries.

Figures from the developing countries themselves confirm those from the developed countries. In the case of Spain, for example, payments for technical assistance in 1979 and 1980 were more than double those for royalty payments. For Turkey, in the same years, payments for technical assistance were many times as great as payments for licences and know-how. Service payments by Turkey connected with external project credits (i.e. expert fees, payments for engineering work, payments for consulting) were however higher than all other items together in Turkey's technological balance of payments.

While transactions with developing countries are relatively small from the point of view of such countries as the United States, the absolute figures seem large to the developing countries themselves. In 1978, for example, US receipts for royalties etc. from these countries amounted to over $ 1,000 million. The size of the payments, and the fact that the bulk of them are made between affiliated firms, has focused a good deal of adverse criticism on them on the part of developing countries. The issue has been raised at successive UNCTAD conferences, and it has proved impossible to eradicate the suspicion that large multinational companies, mainly based in the USA and Europe, have taken advantage of lack of technical and commercial knowledge in developing countries. Indeed, a UN Centre on Transnational Corporations was set up in 1978 partly to help developing countries with these problems.

The most positive action to deal with the problem has been taken by countries in Latin America. Argentina, Brazil, Mexico and the Andean Pact countries all introduced regulations and controls on technology transactions during the 1970s. One of the aims of the controls was to limit international payments for technology, while other aims were to improve the bargaining power of domestic firms, and to "unpackage" the various components of technology agreements.

Whether these measures had an effect is not clear, but during the 1970s receipts from Latin America by Germany, Japan, the UK and the USA slowed down in relation to those from elsewhere (Table 9). The increase in annual receipts from Latin America to these four countries was much lower than the increase from the rest of the world. At the same time, direct investment in Latin America grew at a faster rate than receipts for technology (except in the case of Japan). Income and remitted profits from Latin America also outstripped payments for technology. While the controls may have had the effect of reducing the growth of payments for technology, however, it is not clear that they had the effect of slowing down the overall process of technology transfer. There were continued strong inflows of direct investment into Latin America, and heavy international borrowing, partly used for imports of capital equipment. Possibly direct payments for technology were replaced, at least to some extent, by outflows of profits from US and other countries' subsidiaries.

Explanations for Royalty Payments

The figures for the technological balance of payments that have been discussed in this paper are up to a point self-explanatory. This is true particularly of those which relate to the links between developed and developing countries. The overwhelming surplus in the technological balance of payments in favour of developed countries is only to be expected. The relatively high proportion of developed countries' receipts derived from developing countries is surprising at first sight, but it is less so when the attempts of developing countries over a period of years to industrialise are considered, together with the extent of the involvement of multinational companies in the process of industrialisation.

Substantial payments for technical assistance are made by developing to developed countries in connection with capital projects, and in many cases the figures for royalty payments are dwarfed by these. Such payments are again easy to explain in principle. The level of payment for both royalties and technical assistance has however been questioned in some developing countries, notably in Latin America.

The figures that are more difficult to explain are those for the technological balance of payments between developed countries. At first sight, the large deficits of countries such as West Germany and Japan, and the surplus of the United Kingdom, do not seem to square with what is otherwise known of the industrial progress of these countries. Nor do they square with what is known of research and development expenditure in the countries concerned.

Some relationships concerning research and development expenditure and other variables seem to be better understood than those connected with deficits on the technological balance of payments. For example, the relationship between research and development expenditure in a country, and that country's success in international trade, now seems to be well established. It has recently been shown (Soete 1980) that differences amongst OECD countries in their export shares in forty sectors, covering all manufacturing industry, are best explained statistically by differences among these countries in their research and development activities. It seems to be accepted also that there is a link between the successful assimilation of foreign technology and an accompanying high level of indigenous R & D and investment. This has been found in the case of Japan by Oshima (1973). Pavitt and Soete (1982) have established that this finding has general application. These studies suggest that, to be successful, the import of foreign technology cannot be regarded as a substitute for domestic R & D. It must be built upon by the importing country, and if this is done it can lead to further progress, as we know has been the case in Japan. Where, however, indigenous talent and resources are lacking, much imported technology has gone to waste.

Not surprisingly, exports of technology and direct investment overseas seem to go together. Vickery (1981) has investigated Japanese data on technology exchange and foreign direct investment in recent years. He has shown that Japanese receipts for technology exports in 1979 show a good correlation (R = 0.788) with cumulative foreign direct investment across nine manufacturing industries. In addition, the pattern of Japanese manufactured exports to Asian countries closely follows the pattern of technology exports, and hence of investment. At the same time, Asian countries which have received Japanese technology seem to have been successful in their own exports. There is a high correlation (R = 0.985) between Japanese receipts for technology exports in manufacturing industry to Asia in 1972-79, and imports of manufactured goods from Asia by all OECD countries.

The historical nature of technological balance of payments data has previously been stressed. Some interesting figures released by the Prime Minister's office in Japan (Haine 1981) emphasize this. Receipts for technology exports have been expressed as a proportion of payments for technology imports, for the years 1971 to 1977, and divided into "new" and "continued" programmes. The figures show that the proportion of receipts to payments for continued programmes rose during the period, but even by 1977 had reached only 0.33. For new programmes, on the other hand, the proportion exceeded 1 from 1972 onwards, and reached 2.15 by 1977. In the case of new programmes in iron and steel and chemicals, in particular, the figures were well above these averages.

The nature of royalty agreements suggests that there should be a relationship between payments for technology (at constant prices) and the volume of industrial output, since royalty payments are often based on a proportion of sales revenues. In order to test this relationship, we compared industrial output from 1970 to 1980 in the leading industrial countries (expressing the volume of output in terms of index numbers) with technological payments and receipts (expressed in constant 1975 US dollars, as shown in Tables 1 and 2). The results are given in Table 10. The correlation coefficient (R) between payments and output is high for France and quite high for West Germany. It is also high for the USA, but weak for Italy and the UK. In the cases of Japan and Switzerland it is actually negative, and the R^2 is low.

The figures for Japan in particular are not difficult to explain. Payments declined in real terms during the 1970s, while industrial output rose by well above the OECD average. The weak Italian relationship is due to a fairly flat trend in payments, while output rose over the period in line with the OECD average. The UK figures, as so often, are a thing apart. The behaviour of UK output, although weak, was not out of the ordinary but movements in payments followed no clear pattern.

It can be argued that there is a relationship between industrial output in a given country and receipts for technology. Countries with a strong industrial performance are likely to have a good export performance. Good exports may, as in the case of Japan, go with foreign investment and exports of technology. When the relationship is tested, the correlation between receipts and output actually turns out to be stronger than that between payments and output (Table 10). It is high for France, Japan, Italy and West Germany, and moderately high for the UK and the USA. It is however negative for Switzerland. The figures may be affected by the fact that output in all these countries moved broadly in the same direction during the 1970s, and that their technology receipts from each other form a major part of their total technology receipts, so that their receipts reflect OECD output rather than their own.

It has to be remembered in all these comparisons that, in so far as payments for technology are based on output, the sectors mainly concerned in technology transfer form a part only of total output. In addition, the type of output on which royalties are paid varies over time. It is thus scarcely to be expected that output and expenditure (or output and receipts) should be strongly correlated.

Parents and Subsidiaries

We now turn to a rather fuller examination of the relationship between affiliated firms than has been attempted so far. The difficulty here is that there is little direct evidence on this question, so that infereneces have to be drawn on the basis of incomplete information.

One thing that does seem to be clear is that flows of intercompany payments for intangible items (covering both profits and royalty payments) are substantially distorted by companies' efforts to minimise their global tax payments. Hirschey and Caves (1981) quote Connor (1977) on this point, in relation to his work on Latin America. In Bond's view (Bond 1981), it is possible that direct investment-related royalty and fee receipts may, because of tax considerations, overestimate the value of the technology transferred. She also believes that unaffiliated transfers may under-represent the true value of the technology transferred, because in this case receipts and payments do not include other means of gaining revenue from exports of technology, such as tie-in sales, agreements to purchase goods from licensing firms, and so on. Independent firms may also deliberately undervalue their proprietary knowledge in order to be competitive with other companies who have a subsidiary in the country concerned.

We have quoted evidence from an earlier study (Taylor and Silberston 1973) that licence agreements between subsidiaries and their parents may provide for a royalty on total sales of the subsidiary, and not just for sales arising from licensed products or processes. Such agreements are bound to lead to a higher level of royalty payments than if they were more narrowly drawn. Agreements between independent firms, on the other hand, must normally relate to licensed output only, although, as we have just seen, other payments (not in the royalty figures) might substantially raise the amounts paid in these cases.

Horn (1981) has been quoted as an author who has cast doubt on the level of German payments to affiliated firms. He pointed out that in German manufacturing only some 15% of employment is dominated by foreign companies, while more than three-quarters of German royalty payments abroad go to affiliated firms. These figures in themselves prove nothing, since by no means the whole of German manufacturing industry relies on high technology, and some sectors which do so may rely on German rather than on foreign expertise. More relevant are other figures quoted by Horn of foreign direct investment in Germany in 1980. Some 95% came from industrialised countries, of which the USA accounted for 35%, followed by Switzerland (15%), the Netherlands (12.5%) and the UK (11.5%). The proportion of royalties paid to affiliated firms overseas closely followed their share of direct investment, with figures of 40% for the USA, 16% for Switzerland and 13% for the Netherlands (although the UK received some 2% of royalties only). Looked at in this way, the German royalty payments seem more reasonable. It is interesting, however, that 74% of German direct investment abroad in 1980 was in industrialised countries, but only 7% of receipts came from affiliated firms in those countries: the remaining 68% of receipts from those countries came from unaffiliated firms. Either German parents do not export technology to their subsidiaries, which seems unlikely, or they do not charge them for it. Another possibility is that the investments are too recent to have yet provided much of a royalty yield.

The Italian case (Scarda and Sirilli 1981) has been briefly mentioned. Once again firm evidence is lacking. Scarda and Sirilli found, however, that payments by subsidiaries of foreign multinationals had a much higher value per operation than in the case of independent firms who imported technology, and that royalties (as opposed to fees) accounted for a higher percentage of subsidiaries' total payments. They concluded that "the theory that Multinational Enterprises use the channel of Technological Balance of Payments in order to transfer the remuneration for the risk capital cannot be rejected".

Figures for the UK throw further light on this issue. In 1980, total net profits of UK subsidiaries, due to overseas parent companies, amounted to £ 1,580 million. Dividends paid to parent companies were £ 742 million (47% of net profits). Royalties paid to overseas parents amounted to £ 314 million, and payments for services to £ 254 million. Royalties and payments for services together therefore amounted to £ 568 million, or 77% of the level of dividends.

The bulk of these payments were made to North America and Western Europe. In 1980, North American parents received £ 605 million in dividends, while Western European parents received £ 93 million in dividends. Royalties paid to North American parents were £ 246 million and payments for services £ 186 million. For Western Europe royalties were £ 66 million and services £ 63 million. The totals amounted to 71% of dividends in the case of North America and 138% of dividends in the case of Western Europe.

The size of UK payments for royalties and services, in relation to those for dividends, show clearly how important royalty and service payments are as a means of remitting funds from subsidiaries to parents, both for US parents and for those from Western European countries. It is difficult to avoid the suspicion that the relative size of these flows are the result of policy decisions which have deliberately enhanced the role of royalty and service payments, in relation to that of dividents.

The Significance of Technological Flows

If it is accepted that royalty payments between parents and subsidiaries may be artificially inflated, is it possible to arrive at a realistic measure of how large these flows "ought" reasonably to be? The task is not an easy one, but with appropriate research some results could be achieved. What would be needed would be a study of particular parents and subsidiaries, and the licence agreements that existed between them. Such matters as royalty rates would have to be looked at, together with the basis of calculation of the sales concerned. One would have to ask whether the royalty rates were at the "normal" level for the type of product or process concerned, and whether the sales taken into account related only to patented products or processes, or to some wider class of sales. Hypothetical "reasonable" royalties could then be calculated, and compared with the royalties actually paid.

In order to carry out such calculations, research would have to be undertaken with full co-operation from multinational firms. In the absence of such research or co-operation, it is not easy to find a satisfactory method of estimating the magnitudes that might be involved. One possible way in theory would be to look at the sales of subsidiaries, broken down by industry, and then to take conventional royalty rates for the industries involved. On the initial assumption that the whole of the output in each industry group was linked with patents, it would be possible to calculate national royalty payments and compare them with actual payments. There would be many problems involved in such a calculation. Although the assumption that 100% of output is related to patents is implausible, it is not clear how one is to arrive at the true percentage without detailed research. What is more, the conventional royalty rates applicable to products may not be those applicable to processes, but the nature of these may not be apparent from the figures of sales.

Would it be possible to make such guesses on the basis of published information? The answer is that it would not, in the UK case. Neither sales, earnings nor royalty figures are available by industry group for affiliates of foreign firms in the UK. We can, however, attempt an aggregate calculation. Let us assume that profits after tax represent 10% of sales, and that royalties on sales are on average 5%. In this case, net profits of £ 1,580 million in 1980 (see last section) would have generated royalties of £ 790 million, assuming that all sales attracted royalties. Royalties actually paid in 1980 amounted to £ 313 million and payments for services to £ 254 million. Actual royalties were therefore at about 40% of the level they "should" have been. Taking royalties and service payments together, the figure would have been 72%. The two items together were equivalent to a royalty of 3.6% on our postulated level of sales. This is an average level of royalty (see Taylor and Silberston 1973, p. 123), although the royalties element alone would have amounted to just under 2%.

On the basis of this crude calculation, the royalty and associated figures seem to be on the high side, since the assumption so far is of royalties on 100% of all sales. We must, however, take into account that profits were lower in 1980 than in 1979, on account of the depression, so that it may be unlikely that net profits in 1980 in fact represented as much as 10% of sales. In this case, the estimated royalty rates would be lower than our calculations suggest. Even on 50% of sales, rather than 100%, they might not look especially high.

These are of course very rough guesses, but they do suggest that, at least in the case of UK subsidiaries of overseas firms, any "padding" of the royalty figures may not have been great. If we say that royalties and other payments should "reasonably" have been at half their actual level, we may have made too great an adjustment downwards.

Let us however assume, for the sake of argument, that royalty and service payments between parents and subsidiaries were twice they "ought" to have been. We then have, in the case of UK subsidiaries, an "excess" royalty payment in 1980 of some £ 280 million. This was 18% of the level of net profits (after tax), and 38% of the level of dividends. If this figure had been added to profits before tax, the extra yield in taxes would probably have been relatively modest. In the light of these British figures, therefore, it does not seem that the British government (or parent companies' governments) need be too concerned about the tax-avoidance effects of excessive royalty figures.

Whether this conclusion can be generalised to other countries and other situations cannot be known without further work that we have not yet been able to undertake. One thing that is quite clear, however, is that royalty and associated payments are in total very much smaller than the main constituents of the balance of payments. In the UK case, for example, technological receipts at current prices were £ 440 million in 1980, while exports of goods and services amounted to over £ 47,000 million. Even in the case of the USA, where technological receipts at current prices were nearly $ 7,000 million in 1980, these were equivalent to 3% only of exports of goods and services of $ 221,000 million.

From a statistical point of view, therefore, the figures of the technological balance of payments are dwarfed by the other international trade magnitudes. This should not however blind us to the great actual and potential influence of technology flows on the industrial development of nations. We have in fact examined the importance of such flows in the case of Japan in another essay (Ledić and Silberston 1984).

Conclusions

It has been seen that the technological balance of payments is a partial indicator only of technology flows between countries. In addition, the figures represent to some extent a historical situation and may not accurately reflect current trends. This has clearly been seen in the case of Japan, where new agreements reflect current Japanese strength far more than old agreements. The figures nevertheless show the continued dominance of the USA as a provider of new technology to both developed and developing countries. Most other developed countries, with the exception of Switzerland and (barely) the UK, have a deficit in their technological balance of payments. Much of the US surplus with other countries reflects the importance of overseas subsidiaries of US multinational firms, and these account for the bulk of

US receipts of royalties and fees. Developed countries, even when in deficit with the USA, have large surpluses with developing countries, transferring to them their own, as well as US-derived, technologies.

The developing countries pay substantial amounts to the developed countries, partly owing to the presence of subsidiaries of US, European and Japanese firms. Royalties are comparatively less important, and payments for technical assistance more important, than in the case of developed countries. Adverse criticism of the level of royalty payments, and of other conditions attaching to technology transfer, has led Latin American countries in particular to work together to control the situation. There is some suggestion that they have succeeded in moderating the level of royalty payments, but this may possibly have led to an increase in the proportion of transfers to parent companies in the form of profits.

Those countries which have been most successful in importing technology, such as Japan, have geared much of their own R & D activity to the exploitation of imported technology. Countries making good use of their technology, whether imported or not, have been successful in exporting commodities. Indeed, the most successful exporters of manufactured goods have been the countries which have been net importers of technology.

There does not seem to be a strong positive relationship between levels of industrial output and payments for imported technology, even though royalty payments are usually based mainly on output. This is no doubt partly because of the historical nature of the transactions behind the technological balance of payments, and partly because royalties are paid on part of industrial output only : they are concentrated in the mechanical and electrical engineering industries (including electronics) and in the chemical industry (including pharmaceuticals). In Japan's case, the correlation between output and payments is a negative one, since rising output has gone together with rising R & D, and a greatly lessened dependence on imported technology.

The figures for the technological balance of payments contain many puzzles which need further investigation. West Germany's deficit actually rose during the 1970s, in strong contrast to Japan, as well as France and Italy. Is the German balance likely to improve, now that German direct investment overseas has risen (Steger 1981)? Why did the surpluses of both the USA and the UK rise, against expectations? The figures suggest that the continued strength of the USA as a centre for technical advance should not be underrated. The UK figures may reflect the fact that the UK is still an important centre for research and development of an industrial nature, in spite of the doubts of Pavitt (1981) and others. It is the European centre for the R & D of many US multinationals, as well as of its own large firms, and this may help to account for its continued surplus.

The role of multinational firms is transferring technology, and in securing its adequate exploitation, cannot be ignored. With the growth of direct investment overseas by West Germany and Japan, as well as by the USA and the UK, technology transfer may well become even more common, and take place more rapidly, than in the past. Already certain US studies have suggested that the time lag for innovation transfers from parent companies to subsidiaries has shortened in recent years (Bond 1981). This speeding up may possibly be connected with the increase in international competition, and it may reflect also the increasing use of "offshore" manufacturing by Western countries, as centres such as Taiwan and South Korea have come to combine high levels of technology and productivity with low levels of wages. R & D facilities seem to remain centred in their traditional locations, but the scope for technology transfer would appear to be increasing rather than decreasing.

The suspicion remains that multinational firms prefer subsidiaries to make payments to their parent companies in the form of royalties and similar payments rather than in the form of dividends. This applies both in developed and developing countries. Tax and political considerations may play a part in this policy, if such a policy indeed exists.

We have looked at some of the possible magnitudes involved, in the case of subsidiaries of foreign firms in the UK, but in that particular case any "padding" of royalties there may have been does not appear to have been of great significance in terms of possible tax losses to the government. We cannot say, without further research, how widely this conclusion might apply.

In general, figures for the technological balance of payments are dwarfed by those for the balance of trade and other international flows. Nevertheless, technological transfer may play a highly significant role in the industrial progress of nations. No better example of this can be found than in the case of Japan.

TABLE 1

Technological Balance of Payments

Receipts – US Dollars (mn) 1975

	1970	1971	1972	1973	1974	1975	1976	1977	1978	1979	1980
France	301	292	351	382	498	459	537	600	616	587	552
West Germany	261	283	331	276	309	324	301	313	337	337	366
Italy	–	–	63	58	89	72	86	145	109	125	136
Japan	–	148	220	239	215	226	256	265	335	353	390
Switzerland	–	–	1489	–	1522	–	1897	–	2013	–	–
United Kingdom	467	442	471	536	549	493	636	596	589	546	500
United States	3233	3359	3511	3867	4186	5300	4140	4229	4921	4650	4549

Source : Mainly based on papers prepared for OECD (1981). The figures were expressed in French francs etc. at 1975 prices, and then converted into US dollars at the 1975 exchange rate between the French franc etc. and the US dollar.

TABLE 2

Technological Balance of Payments

Expenditure — U.S. Dollars (mn) 1975

	1970	1971	1972	1973	1974	1975	1976	1977	1978	1979	1980
France	380	429	534	555	607	550	679	656	664	685	661
West Germany	705	769	773	767	753	834	799	868	871	883	872
Italy	–	–	368	377	325	385	346	414	551	383	386
Japan	–	740	912	812	602	569	545	542	528	568	589
Switzerland	–	–	391	–	469	–	533	–	561	–	–
United Kingdom	436	411	429	458	489	484	504	489	496	424	402
United States	312	318	372	462	379	473	458	389	511	566	497

Source : Mainly based on papers prepared for OECD (1981). The figures were expressed in French francs etc. at 1975 prices, and then converted into US dollars at the 1975 exchange rate between the French franc etc. and the US dollar.

TABLE 3

Receipts less Expenditure — US Dollars (mn) 1975

(Table 1 minus Table 2)

	1970	1971	1972	1973	1974	1975	1976	1977	1978	1979	1980
France	-79	-137	-183	-173	-109	-91	-142	-56	-48	-98	-109
West Germany	-444	-486	-442	-491	-444	-510	-498	-555	-534	-546	-506
Italy	0	0	-305	-319	-236	-313	-260	-269	-442	-258	-250
Japan	0	-592	-692	-573	-387	-343	-289	-277	-193	-215	-199
Switzerland	0	0	1098	0	1053	0	1364	0	1452	0	0
United Kingdom	31	31	42	78	60	9	132	107	93	122	98
United States	2921	3041	3139	3405	3807	4827	3682	3840	4410	4084	4052

TABLE 4

OECD Industrial Production

Index numbers, 1975 = 100

	1970	1971	1972	1973	1974	1975	1976	1977	1978	1979	1980	1981
Australia	91	95	95	105	108	100	106	104	105	114	114	117
Austria	–	–	–	–	–	–	–	–	114	122	125	124
Belgium	93	94	100	107	112	100	109	109	111	116	115	112
Canada	81	87	94	103	106	100	105	108	112	118	116	117
Finland	–	–	–	–	–	–	–	–	–	118	129	132
France	89	93	99	105	108	100	109	110	113	118	118	117
West Germany	96	97	101	108	106	100	107	110	112	117	117	116
Italy	92	92	96	105	110	10)	112	112	114	121	128	125
Japan	91	94	101	116	112	100	111	116	123	133	142	146
Netherlands	84	89	93	100	105	100	106	107	108	112	112	110
Portugal	–	–	–	–	–	–	–	–	125	134	141	142
Spain	71	74	86	99	107	100	106	119	122	114	116	114
Sweden	87	88	90	97	102	100	98	95	93	98	99	95
Switzerland	105	107	109	115	117	100	102	106	107	108	114	114
United Kingdom	97	97	99	108	105	100	103	108	112	116	108	104
United States	91	93	102	110	110	100	111	117	124	129	125	128
OECD Total	91	93	99	108	109	100	109	113	117	123	123	124

Source : OECD, Economic Outlook.

TABLE 5

UK Overseas Royalties and Similar Transactions, 1980
(£ million)

	Receipts			Expenditure		
	Affiliated concerns	Unaffiliated concerns	Total	Affiliated concerns	Unaffiliated concerns	Total
Europe : EEC	53	31	84	34	10	43
: EFTA	7	5	12	24	3	27
North America	80	46	126	224	29	253
Commonwealth	34	12	46	13	3	16
Developing countries*	29	42	71	—	4	4
Oil-exporting countries	4	2	6	–	—	—
Wold Total	214	165	379	283	48	331
of which :						
printed etc. royalties	–	—	60	–	–	34
technological and mineral royalties	—	—	319	–	–	297
World Total**	230	287	517	324	74	398

Source : Business Statistics Office (1980).

* Other Western Europe, rest of world (excluding USSR and Eastern European centrally-planned economies).

** Includes estimates for firms not making reports.

TABLE 6

US Receipts and Payments of Royalties and Fees, 1970-78

(£ million)

1) All receipts and payments

	1970	1971	1972	1973	1974	1975	1976	1977	1978
Total net receipts	2134	2375	2566	3021	3584	4008	4084	4474	5429
Developed countries	1651	1856	2031	2421	2857	3177	3273	3639	4381
of which UK	273	306	334	376	427	523	520	558	742
EEC*	511	585	631	794	954	1091	1043	1187	1466
Japan	268	306	342	426	439	419	485	554	744
Developing countries	483	520	536	599	726	831	813	837	1045
Total net payments	225	241	294	385	346	473	482	434	610
Developed countries	216	234	289	376	346	452	451	481	574
of which UK	54	48	59	73	84	103	85	91	159
EEC*	54	58	63	95	75	84	92	100	176
Japan	8	5	7	14	−35	−17	−21	−18	−5
Developing countries	9	7	6	10	−1	20	33	16	36

TABLE 6 Ctd

TABLE 6 Ctd

US Receipts and Payments of Royalties and Fees, 1970-78

(£ million)

2) Receipts and payments with affiliated firms

	1970	1971	1972	1973	1974	1975	1976	1977	1978
Total net receipts	1561	1757	1911	2309	2833	3251	3262	3554	4364
Developed countries	1142	1299	1448	1783	2200	2522	2570	2849	3466
of which UK	217	240	271	302	356	444	448	476	649
EEC*	354	424	473	625	767	892	833	961	1205
Japan	66	83	102	153	190	200	239	279	401
Developing countries	418	458	464	525	632	729	693	707	896
Total net payments	111	118	155	209	160	287	293	243	396
Developed countries	108	115	154	208	167	270	267	237	374
of which UK	19	11	15	20	17	27	8	19	75
EEC*	2	3	6	23	5	17	25	37	111
Japan	4	1	1	1	-47	-26	-34	-34	-66
Developing countries	2	3	1	1	-9	16	27	4	23

Sources: Bond (1981) and US Department of Commerce, Survey of Current Business

* Original six members only

TABLE 7

Germany : Receipts and Expenditure on Patents. Inventions etc.
(Per cent)

	Industrial countries		Developing countries		Centrally-planned economies	
	Receipts	Expenditure	Receipts	Expenditure	Receipts	Expenditure
1970	70.9	99.3	27.2	0.4	1.9	0.3
1971	67.6	99.3	29.8	0.4	2.6	0.3
1972	69.8	99.5	25.8	0.5	4.4	0.0
1973	72.2	99.6	24.6	0.3	3.2	0.1
...						
...						
1976	74.7	99.0	19.1	0.7	6.2	0.2
1977	72.6	99.0	18.2	0.5	9.2	0.5
1978	75.4	99.4	19.3	0.4	5.3	0.2
1979	73.8	99.5	21.6	0.0	4.6	0.2
of which affiliated firms						
1979	7.2	77.5	1.6	0.0	0.2	0.1

Source : Horn (1981).

TABLE 8

Receipts for Technical Assistance and Related Services and for Industrial Property Rights
(percentage of totals)

	Developed countries	Developing countries
France (1980)		
Patents and licences	80	20
Technical assistance	67	33
Italy (1978)		
Patents, licences, trademarks, inventions	85	15
Technical assistance and know-how	76	24
Non-connected technical assistance	62	38
United States (1980)		
Affiliated :		
Royalties and licence fees	92	8
Service charges and rentals*	57	43
Unaffiliated :		
(mostly royalties and licence fees)	82	18

Source : Vickery (1981).

* Includes fees for management and technical services charged by parents to affiliates.

TABLE 9

Growth in Receipts for Technology from Latin America Compared with Rest of World
(million current US $)

	Latin America						Rest of world % change
	1970	1971	1972	1979	1980	% change	
West Germany	21.4			39.0		82	363
Japan (manufacturing)			6.9	32.9		377	760
United Kingdom		9.2		16.8		83	121
United States :							
Affiliated	241				566	135	269
Unaffiliated	45				101	124	102
For comparison : Direct Investment Stock							
West Germany	843			4235		402	
Japan			640	2583		304	
United Kingdom		1270		2900		128	
United States	12961				38275	195	

Source : Vickery (1981).

TABLE 10
Correlation Results

1) Expenditure (Table 2) and Industrial Output (Table 4)

	R	R²	N
France	.9490	.9006	11
West Germany	.7830	.6131	11
Italy	.1675	.0281	9
Japan	− .4392	.1929	10
Switzerland	− .4332	.1877	4
United Kingdom	.1364	.0186	11
USA	.8212	.6743	11

2) Receipts (Table 1) and Industrial Output (Table 4)

	R	R²	N
France	.8992	.8086	11
West Germany	.6800	.4624	11
Italy	.8053	.6486	9
Japan	.9455	.8940	10
Switzerland	− .6875	.4727	4
United Kingdom	.5909	.3491	11
USA	.6106	.3728	11

REFERENCES

BOND, J.S. (1981), "United States International Transactions in Royalties and Fees — Trends and Interpretation", Workshop on the Technological Balance of Payments, OECD, 14th-15th December 1981.

BUSINESS STATISTICS OFFICE (1980), *Overseas Transactions*, Business Monitor MA4, HMSO, UK.

HAINE, I.R. (1981), "The Technological Balance of Payments — Australian Statistics", Workshop on the Technological Balance of Payments, OECD, 14th-15th December.

HIRSCHEY, R.C. and CAVES, R.E. (1981), "Research and Transfer of Technology by Multinational Enterprises", *Oxford Bulletin of Economics and Statistics*, May.

HORN, E.-J. (1981), "Technological Balance of Payments and International Competitiveness : The Case of West Germany," Workshop on the Technological Balance of Payments, OECD, 14th-15th December.

LEDIĆ, M. and SILBERSTON, A. (1983), "Some Aspects of the Technological Balance of Payments". Paper read to Sesquicentennial Conference of the Manchester Statistical Society, April.

—— and —— (1984). "The Technological Balance of Payments, Technical Progress and International Trade". Paper read to Fifth Sewanee Economics Symposium, March (to be published).

OECD *Economic Outlook* (monthly).

—— (1981), Workshop on the Technological Balance of Payments, 14th-15th December — see separate papers.

OSHIMA, K. (1973) in Williams. B. (ed.) *Science and Technology in Economic Growth*, MacMillan.

PAVITT, K. (1981) "Technology in British Industry : a Suitable Case for Improvement" in Charles Carter (ed.), *Industrial Policy and Innovation* Heinemann.

—— and SOETE, L. (1982), "International Differences in Economic Growth and the International Location of Innovation" in Giersch H. (ed.), *Emerging Technology : Consequences for Economic Growth, Structural Change and Employment in Advanced Open Economies*, J.C.B. Mohr (Paul Siebeck) : Tubingen.

SCARDA, A.M. and SIRILLI, G. (1981), "Technology Transfer and the Technological Balance of Payments (Italy)", Workshop on the Technological Balance of Payments, OECD, 14th-15th December.

SCHIFFEL, D. and KITTI, C. (1978), "Rates of Invention : International Patent Comparisons". *Research Policy* 7.

SOETE, L. (1980), *The Impact of Technological Innovation on International Trade Patterns : The Evidence Reconsidered*, OECD, STIC /80.33, Paris, August.

STEGER, A. (1981), "The Recording of Patent and Licence Transactions in the German Balance of Payments", Workshop on the Technological Balance of Payments, OECD, 14th-15th December.

TAYLOR, C.T. and SILBERSTON, A. (1973), *The Economic Impact of the Patent System*, Cambridge University Press.

UNITED STATES DEPARTMENT OF COMMERCE, *Survey of Current Business* (Montly).

VICKERY, G. (1981), "International Technology Transactions: Data and Interpretation", Workshop on the Technological Balance of Payments, OECD, 14th-15th December.

CHAPTER SIX

NEW TECHNOLOGY REVOLUTION:
MYTH OR REALITY FOR DEVELOPING COUNTRIES?

By A. S. Bhalla* and J. James**

I. *Introduction*

Technological backwardness is a major distinguishing feature of under-development. A very large number of people in developing countries, especially those living in poverty, are dependent on traditional technologies that are incapable of generating levels of income adequate to meet even the most basic human needs. Without a sharp increase in the productivity of these technologies, it is doubtful that any significant reduction in global poverty can be effected.

It is in the light of this vast and pressing problem that we think the relevance of the new and emerging technologies in the industrialised countries ought to be appraised. In particular, if this "New Technology Revolution" is to become anything like a reality for the developing countries it will need to produce a beneficial impact on the productivity and earnings of the vast majority of the inhabitants of these countries. What this in turn, requires, is that the new and emerging technologies be applied to upgrading the abysmally low productivity of the traditional technologies which, as noted above, underlies much of the poverty in the Third World. This essay, accordingly, is concerned to assess the scope and prospects for the integration of emerging and traditional technologies and to examine ways in which the likelihood of successful integration can be increased.

Our first task is to analyse the notion of integration and to specify the requirements for successful integration.

In this essay, we distinguish the following four groups of potential users of traditional and emerging technologies :

* International Labour Office, Geneva.

** Boston University, U.S.A.

This essay is a revised and enlarged version of an earlier paper "An Approach Towards Integration of Emerging and Traditional Technologies", ILO World Employment Programme, Research Working Paper No. WEP 2-22/WP. 112, Geneva, February 1983. The views expressed in the essay are our own and not necessarily shared by the organisation.

A. *Rural*

1. Traditional agriculture and rural non-farm activities
2. Rural cooperatives

B. *Urban*

3. Urban small-scale producers (small-scale organised sector and urban informal activities)
4. Large-scale industrial sector.

These four groups of users are distinguished by a number of criteria, namely, scale of production, social and economic organisation and work patterns, modes of employment (self-employment or wage-labour), nature and quality of products, and market and income characteristics, etc. The capacity of each of these potential users to effectively absorb new technologies is examined in the following pages (where we also specify the sense in which the large-scale industrial sector embodies traditional technologies).

The process of integration or blending of new with traditional technologies can be interpreted in two distinct ways, viz.

(a) *Coexistence* of new and traditional or old technologies in a given area and economic activity (there may be no resulting change in the character of traditional technologies);

(b) *Integration* of the two which leads to an improvement of traditional technologies, in terms of (i) cost per unit, (ii) productivity per unit of factor input, and (iii) quantity and quality of output[1].

In addition to these cases, there is also a real situation, in developing countries, of *disintegration* which may be defined as the displacement of traditional technologies by the new technologies over time. The experience of many developing countries shows that the blacksmiths, carpenters and similar other craftsmen have gradually disappeared in the face of competition from modern cost-reducing technology.

II. *The Requirements of Successful Integration*

In this section, we describe the requirements for successful integration of traditional and emerging technologies in relation to each of these groups and highlight the major respects in which these requirements may be ex-

---

1. There is no reason to confine the notion of integration to the application of one particular kind of new technology at a time. Possibilities of integrating different kinds of new technologies should not be overlooked.

pected to differ across them We then apply the criteria for successful integration to the emerging technologies in order to identify those among them that appear to be the most promising candidates for further research and development.

The most meaningful concept of integration could be defined in terms of a fusion of traditional and emerging technologies into a new entity. However, even this type of integration can take a wide variety of forms depending upon the weighting of the two types of technologies in the fusion process. At the one extreme, the new entity will embody almost entirely the characteristics of the traditional technology and at the other, it will reflect mainly the elements of the new or emerging technology. As a general rule, if integration is to be successful, it should not be of the latter kind (though of course, there may be exceptions); that is, the outcome of the integration process should not be too far from the traditional technology (in a sense to be more fully explained below). In the case of traditional producers in rural areas the following specific reasons for this requirement may be cited.

1. Rural Sector

(a) Traditional Producers

Here we deal only with traditional farmers because this is clearly the dominant group. It is necessary, however, to recognise that there are diverse groups of non-farm producers in the rural sector — such as cobblers, blacksmiths, carpenters and other craftsmen — whose traditional technologies are also often sorely in need of upgrading because they result in incomes that are very low.

The requirements for successful application in the case of traditional farmers (who comprise much of the Third World's agriculture) derive from their special socio-economic characteristics. Most importantly, these farmers are usually very poor; often even in relation to average income in the developing country. Indeed, in some countries traditional producers in the rural areas comprise a relatively high percentage of those classified as living in poverty. In many, if not most cases, moreover, they are illiterate and produce mainly for subsistence needs rather than sale in the market.

Finally, we need to mention the paucity of infrastructural facilities (such as irrigation, repair and maintenance, etc.) that are typically available to this kind of traditional farming.

The importance of the above characteristics of peasant farmers is that they impose highly stringent requirements for successful integration of tra-

ditional and emerging technologies. Firstly, the innovation must be close
to the traditional practice in the sense that it must not demand cash outlays
that are large relative to the farmer's highly limited resources. Secondly,
and relatedly, the innovation must not impose a high degree of risk on the
farmer. Because he produces mainly for subsistence and earns only a limited
amount of cash from market sales, the farmer is understandably reluctant
to venture these resources for new techniques of production whose outcome
is by no means certain and may even be hazardous. Thirdly, the new technique must be comprehensible to the peasant farmer. That is, it must not pre-
suppose knowledge that he does not (and often cannot) possess and it must
be recognised by him as "do-able". This condition is more likely to be satisfied
the more closely the innovation fits in with the prevailing ecological system
(e.g. the pattern of local resource use, waste collection, etc.)[2]. Finally, the
new technique must not pre-suppose the existence of facilities – for power,
transport, repair and maintenance and so on – that are simply not available
to traditional producers.

So far we have been discussing only the *demand* side of the requirements
of successful integration with respect to small farmers. That is, we have confined our analysis to the requirements of integrated technologies if they are
to be demanded by this group. But the special characteristics of peasant farming also pose requirements on the *supply* side. What is important to recognise here is that the mere supply of integrated technologies is not sufficient – the fact that these techniques exist and information about them also
has to be communicated to the traditional producers. And the location of
these farmers (who as noted above are frequently illiterate) in scattered and
isolated rural localities often makes this a formidable problem.

(b) Co-operative Forms of Rural Production

A stringent set of requirements for the successful integration of traditional technologies in rural areas and the technologies that are now emerging
(noted above) is based on the assumption that traditional producers are
organised along essentially individualistic lines. One way in which the strin-
gency of the requirements may be relaxed is the organisation of production
on a co-operative basis which allows the resources of individual farmers and
artisans to be pooled for the purposes of communal investment in productive
equipment, inputs and infrastructural facilities. To this extent, it becomes
possible to exploit some of the technologies which, from the standpoint of

2. For a discussion see National Academy of Sciences (1982).

the individual producer, would be inaccessible. Co-operatives appear to have been successful in a number of countries. It would be a mistake, however, to conclude that co-operatives are a panacea for all the difficult problems that are associated with the application of new technologies in rural areas. Particularly in countries without a strong communal ethic, these types of organisations can often only be implemented with considerable difficulty and many have failed. (More on this in Section VII below). The most that can therefore be concluded is that co-operatives offer one possible way of relaxing the severe constraints on the application of emerging technologies to traditional activities that are imposed by the nature of traditional agriculture. Another is "subcontracting" by large industrial enterprises to small and "cottage" (rural) producers. This latter case is examined in the section below.

2. Industrial Sector

(c) Small-scale Producers

Many producers in the small-scale urban sector share the characteristics of peasant farmers and rural craftsmen that we described above. In particular, many of these producers live in absolute poverty, are self-employed and uneducated or school drop-outs.

One of the main distinctive features of the urban small-scale sector is its capacity to generate income-earning employment opportunities. While there is no doubt ample scope exists for improving traditional technologies used and the products produced in this sector through marginal inputs of capital and skills, it is difficult to envisage that these producers can have easy access to such emerging technologies as microelectronics which are developed in and for firms in rich countries. To quote King (1974) on the informal machine-makers in Kenya, power-driven imported machinery (whose price ranges from 3,000 to 10,000 K.sh) is generally "beyond the range of the enterprising and successful workers in the informal sector of the economy. And this is not only that they do not own workshops to which electricity can be brought (indeed there are a sizeable number of artisans who have power available in their premises but still cannot afford to put anything at the end of the cables)"[3].

3. See King (1974), pp. 24-25. The urban small-scale sector includes what is often referred to as the informal sector (typically based on self-employment, unsophisticated and smallscale production) as well as more modern firms which produce on the basis of wage-labour on a somewhat larger scale.

However, one way in which applications of emerging technologies in the urban small-scale sector seem feasible is through the formal (large-scale sector subcontracting some work to the former, and providing technology, skills and other types of assistance. Organisation of small-scale producers (e.g. knitters and weavers) in cluster-type workshops might for example facilitate an economic use of simple numerically-controlled machine tools in the manufacture of some consumer goods. (See discussion of the Italian Prato Textile Industry below). Nevertheless such applications, even if they can be successfully organised may be socially undesirable if they displace employment in the sector.

Small-scale producers in the industrial sector are differentiated from producers in the urban informal sector in a number of respects that are relevant to discussing the requirements for the successful integration of disparate technologies.

At the risk of over-simplification, the former can be said to be differentiable from the latter by virtue of more closely resembling the model of the firm in rich countries. That is, small-scale producers in industry tend to operate according to the principles of profit-maximisation on the basis of wage labour. Compared to most firms in rich countries, however, they produce on a very small-scale using relatively old technology. Though their incomes are usually very low in relation to average earnings in developed countries, small-scale producers in the industrial sector are rarely among those classified as living in poverty in the Third World.

(d) Large-scale Industrial Firms

It may seem surprising that we include large-scale firms as the final source of traditional technologies for it is well-established that the size of firm in a given industry is positively correlated with the degree of modernity of the technology that it employs. However, since some production processes (or sub-processes) are much more resistant to technical change than others, this general association does not exclude the existence of traditional techniques in *some* aspects of the large firms' behaviour. (Equally, it is often possible to find elements of modernity within the production processes of small firms in the industrial sector) For example, transport functions within large firms may sometimes be conducted manually rather than with the use of a conveyor belt. It is these, more traditional aspects of the production of large firms that offer the promise of integration with the emerging technologies.

III. *A Matrix Approach to Integration*[4]

The emerging technologies may be classified into three main areas, namely, micro-electronics, biotechnology and renewable energy technology. Since we have already identified four main groups engaged in traditional activities it is clear that there are twelve separate integration possibilities. In what follows we try to fill in these twelve cells, of what in effect is a matrix approach, with concrete examples (see Table 4 below).

Drawing on the discussion in the previous pages, the following two sections attempt to assess the prospects for successful blending of these technologies with those used by the four groups of agents noted above. This exercise makes no pretence to being exhaustive in coverage or to being conclusive. Rather, what is intended is merely an attempt to focus the debate on what seem to be the most promising directions for the future.

The following discussion shows that some new technologies, e.g. microelectronics and computer systems are becoming a reality in both the industrialised and developing countries. The microelectronics are sufficiently developed and diffused to be availed of in agriculture and industry. This does not seem to be the case with biotechnology and genetic engineering where much R and D still remains to be done before the technology can be applied effectively to solve the food, nutrition and other problems of developing countries. The case of renewable energy technologies, like flat-plate collectors and photovoltaic cells, seems to be somewhere in between the other two : although some scattered applications are already noticed in developing countries, new solar technology has not yet achieved any major breakthrough.

In the following sections, wherever appropriate we shall draw a distinction between new technologies that are already being applied in developed and developing countries and others whose application at present is within the realm of speculation and uncertainty. This distinction has practical implications for determining future research priorities and for policy-making.

1. *Potential for Application in the Rural Sector*

Of the three emerging technologies considered in this essay, the two, namely, biotechnology and renewable energy sources seem particularly promising for agriculture. However, the applications of microelectronics to agricultural operations seem to be more limited at present than their applications to industrial products and processes.

4. This section is based largely on Bhalla and James (1983).

A. *Microelectronics*

On cost grounds alone the prospects of successful integration in the field of microelectronics seem limited to co-operative forms of organisation in rural areas and small-scale producers located in urban areas.

According to Radnor (1982) electronic materials can cost anywhere between $ 20 and $ 200 for low volumes (i.e. volumes that go as low as 30-50 units per year)[5]. To these figures have to be added the cost of non-electronics components (e.g. housing) and also any needed sensors. Taking all these costs together, it seems not unreasonable to suppose that the overall figure could easily exceed $ 300, an amount which is greater than annual per capita income in India and other poor countries.

There *may*, however, be greater opportunities for application of emerging technologies within the framework of rural co-operatives and small-scale producers. For example, there could be opportunities in :

(i)   food storage control (microelectronic control can ensure automatic shifting of grain, reducing spoilage and use of energy);

(ii) moisture control (microprocessor-controlled testers can be used to pre-test and control moisture content in food-grains);

(iii) sprinkler control in irrigation (microprocessors can be used to regulate the timing and flow rate of water).

The above cases are however only potential possibilities which have not yet been realised in many developing countries on any significant scale. One microprocessor device that is said to be already in use in some developing countries is that for measuring the fat and solid non-fatty content of milk. This device is said to have "reduced the cost of these measurements considerably, is non-destructive and hence saves large volumes of milk, and can be operated with a small car battery, making it relevant for decentralised rural use (e.g. village co-operatives)"[6].

Village co-operatives will seemingly also be required to take advantage of the three potential possibilities noted above.

B. *Biotechnology*[7]

The vast majority of traditional technologies are located in the rural

5. See Radnor (1982).
6. See Radnor and Wad (1981).
7. For a general discussion of biotechnology, see Ventura (1982), pp. 109-29, and Héden (1979).

sector of developing countries. Because several of the emerging biotechnologies have applications in this sector, they offer considerable scope for the integration of HYV technologies with the new innovations which can accelerate the development of renewable agricultural resources available within the developing countries[8]. For example, the benefits of the Green Revolution in many parts of the Third World could not be fully tapped for lack of adequate fertilisers and irrigation inputs. Gene splicing (genetic engineering) could offer new ways to provide varieties of plants and crops that can make their own fertilisers. Natural nitrogenous-fixing bacteria would minimise the need for costly petroleum-based chemical fertilisers. The poor farmers who at present have no access to the expensive fertilisers could hope to realise crop yield increases through the use of such new innovations.

Similarly, new tissue culture which enables single plant cells to grow into complete plants, permits much faster propagation rates than those that can be attained through conventional techniques like seeding, cutting and grafting. Tissue culture methods could accelerate the process of selection of improved tree crops that could be comparable with the impact of the Green Revolution on the cultivation of cereals. It is reported that "clonal oil palms" are already being field-tested on plantations in Malaysia. One commercial company expects, by the mid-eighties to be marketing clones with a proven capacity for high yields"[9].

A combination of different types of biotechnologies holds out much promise to increase crop yields, produce hardier plants, extend growing seasons, strengthen resistence to pests, diseases, heat, frost, drought and flooding, and increase the ratio of edible material to waste matter[10]. Another area of future promise includes developments in mono-clonal antibody research that might lead to production of vaccines against many important human and animal diseases.

With the advancement in biotechnology, the so-called "fermentation industry" has been able to utilise agricultural waste materials to produce alcohols, organic acids, vitamins, vaccines and fodder protein (single cell protein). By using the single cell protein (SCP) instead of grain in animal feeds (and also for direct human consumption) vast amounts of grains and legumes, hitherto consumed by animals and human beings, can be released.

The price of SCP compared to conventional protein (e.g. milk, eggs,

8. World Bank (1982), p. 64.
9. Narang (1981).
10. See National Academy of Sciences (1976), Appendix B on "Single Cell Protein: Its Status and Future Implications in World Food Supply".

meat) is estimated to be very low. It is further estimated that "one SCP plant making 100,000 tons per year can produce about as much protein as that which could be extracted from 120,000 hectares (300,000 acres) of soybeans, or as much beef (cattle) as could be grown on 2 million hectares (5 million acres) of grazing land"[11]. The rising prices of non-microbial protein and of oil have further given a boost to the fermentation industry. In Japan, for example, this industry accounts for 6 per cent of the national income[12].

At this stage the commercial viability of SCP is not at all clear especially in view of rising petroleum prices and high energy requirements for production. What does seem clear though is that "SCP is not a panacea that by virtue of its greater efficiency will displace conventional agricultural and animal sources of protein. The methods of the past will continue to be needed, despite their relative inefficiencies... Improvement of the old ways and adoption of the new will go hand in hand, complementing each other"[13].

In the developing countries, existing processes like sewage farming and fish culture in waste ponds can be made hygienically safe. The use of wastes in the acquaculture of algae or plants is currently being explored. The following is an illustrative list of waste treatment processes which employ microorganisms to yield valuable products :

— production of fodder yeast on spent liquor from paper manufacture by the sulfite-pulping process;

— conversion of carbohydrate wastes into rich animal feeds;[14]

— mushroom production of rice straw or compost;

— methane fuel formed by anaerobic digestion of animal manure or plant residues.

Successful application of biotechnologies, however, will not come about automatically or in the very short run. It will require careful planning and considerable research to ensure that the stringent set of requirements specified earlier are fulfilled. Table 1 below indicates that the average time required to implement genetic production varies from 5 to 15 years.

Apart from the positive effects, the biotechnology research in the advanced countries, and the consequential development of derived products

11. Héden (1979), *op. cit.*

12. National Academy of Sciences (1976), Appendix B, *op. cit.*

13. A citrus processing factory in Belize (Central America) has fermented the citrus fruit with a mould to obtain a product containing 20-25 per cent protein suitable for ruminating feeding. See World Bank (1981).

14. Narang (1981), *op. cit.*

TABLE 1

Market Predictions for Implementation in Production of Genetic Engineering Procedures

Product category	Number of compounds	Current market value in million $	Selected compound or use	Time needed to implement genetic production (years)
Amino Acids	9	1,703	Glutamate	5
			Tryptophan	5
Vitamins	6	667.7	Vitamin C	10
			Vitamin E	15
Enzymes	11	217.7	Pepsin	5
Steroid Hormones	6	367.8	Certisone	10
Peptide Hormones	9	268.7	Human Growth Hormone	5
			Insulin	5
Viral Antigens	9	n.a.	Foot-and-mouth Disease Virus	5
			Influenza Viruses	10
Short Peptides	2	4.4	Aspartame	5
Miscellaneous Proteins	2	300	Interferon	5
Antibiotics	4*	4,240	Penicillins	10
			Erythromycins	10
Pesticides	2*	100	Microbial Aromatics	10
Methane	1	12,572	Methane	10
Aliphatics (other than methane)	24	2,737.5	Ethanol	5
			Ethylene Glycol	5
			Propylene Glycol	10
			Isobutylene	10
Aromatics	10	1,250.5	Aspirin	5
			Phenol	10
Inorganics	2	2,681	Hydrogen	15
			Ammonia	15
Mineral Leaching	5	n.a.	Uranium	
			Cobalt	
			Iron	
Biodegradation	n.a.	n.a.	Removal of Organic Phosphates	

n.a. = Not available.

* = Number indicates classes of compounds rather than number of compounds.

Sources: US Congress, Office of Technology Assessment: Genex Corporation in *Industry Week*, September 7, 1981, p. 68 as shown in Alan T. Bull, Geoffrey Holt and Malcolm D. Lilly: *Biotechnology : International Trends and Perspectives* (Paris, OECD, 1982, p. 73).

like enzyme-based sugar syrups and fructose sweeteners from maize, is likely to threaten the economies of developing countries producing and exporting sugar cane and sugar beet.

Finally, the degree to which the potential of biotechnology is actually realised for the benefit of the developing countries' producers depends on the cost at which these techniques can be made available to poor farmers and small producers which in turn is partly a function of the amount of research that is devoted to this specific problem. It also depends on the way in which rural production is organised — cooperative forms will generally enlarge the extent to which the benefits of biotechnology can be exploited in the future.

C.   *Solar Energy Technologies* (SET)

Some forms of solar technologies appear to be suited for application in the traditional rural activities. In this context, the example of flat-plate collectors is sometimes quoted since they are claimed to be relatively cheap (although cost still appears to be a major factor limiting the application of this technology to developing countries), simple to operate and capable of decentralised use.

The instalment of photovoltaic power technology in a remote part of Upper Volta as part of an AID programme in 1979 illustrates the potential for the application of solar technologies in even the most traditional rural areas. The photovoltaic system powers a grain mill (producing enough to meet the daily requirements of about 640 people) and a water pump[15]. It appears that the villagers themselves were able to carry out some of the installation of the power system and also to cope with its operation and maintenance. In February 1983, National Aeronautics and Space Administration (NASA) Lewis Research Centre installed four stand-alone photovoltaic systems in a village in Tunisia to supply power to domestic, commercial and public sectors. Photovoltaic power systems may also be useful in rural cottage industry with respect to, for example, metal-working and wood-forming. However, like flat-plate collectors, it appears that in 1983 the costs of photovoltaic equipment have not decreased as much as predicted in most of the scientific literature.

15. See Bifano *et al.* (1979).

2. Potential for Application in the Industrial Sector

In industry in developing countries microelectronics seem to have much greater scope for successful applications than either biotechnology or solar energy technologies — at least in the foreseeable future.

A. Microelectronics

In principle, microelectronics should offer the possibility of cost reductions and increased product quality for both small-scale and large-scale producers in the industrial sector. One of the major reasons for expecting microelectronic applications to grow, presumably in decentralised minifactories, is the assumption about economic feasibility of small batch production.

As we have stressed throughout, the possibilities of integration depend on the way in which traditional production can be organised so as to exploit the advantages of the new technologies. We argued that in agriculture this could be achieved through the formation of rural cooperatives. In the industrial sector a similar means of expanding the possibilities of integration would be the encouragement of subcontracting relationships between large and small firms. Such arrangements would, by allowing the latter access to the skills and infrastructural facilities of the former, reduce the risk of adopting the microelectronics technologies which, as we argue below, are highly dependent on a range of complementary inputs for their efficient operation. As a means of upgrading the technological capabilities of small firms subcontracting proved highly valuable in the Japanese case[16] and we believe that there are important lessons to be drawn from that experience in the context of integration possibilities in the industrial sector (see Section VII below).

Microelectronics can perform a number of functions in industrial applications. A distinction can be usefully made between these applications in (i) production and (ii) management, accounting and marketing. Microelectronic applications in direct productive process are mainly of the following type: control of movement of new materials, component parts, finished product, etc.; shaping, designing, cutting, mixing and moulding of materials; assembly of components into sub-assemblies and finished products; product quality control through inspection, testing and analysis; design, manufacture and maintenance operations[17]. A number of computer applications have been

16. For a full discussion, see Watanabe (1983).
17. See Bessant et al. (1981), p. 199.

noted in : stock-keeping, inventory control; allocation of tasks and ordering and handling of materials; cost accounting and invoicing and other financial data collection; and job control and personnel data collection.

In general, there seems to be greater immediate scope and application of microelectronics and other devices in management, materials-handling and marketing functions, than in manufacturing proper. This is partly because the latter requires greater skills in the handling of microelectronic technology than the former. Moreover, the applications to management are less likely to displace jobs and more likely to raise efficiency through waste reduction, faster deliveries, etc.

The example of Prato textile industry in Italy described below shows that the use of computers is much more noticeable in marketing, banking, information exchange, etc. than in the fabrication of textiles. This also seems to be the experience of developing countries. For example, in *China* computer applications have been reported in the management of finished products in the Shanghai (No. 1) Printing and Dyeing Mill. Prior to computer applications, manual methods resulted in an unnecessary high stock-piling, and a high wastage rate of 1 per cent per annum, or a loss of one million metres of cloth. In June 1981, the Shanghai plant established two microcomputer systems with the collaboration of the Huadong Institute of Textile Engineering. The introduction of the computer system was designed for quality control and forecasting of export markets. It is recorded that the new technology applications have reduced the wastage rate from 1 per cent to 0.1 per cent or savings of 900,000 metres of cloth per annum[18].

In *India*, a small firm, has introduced computers for "handling bulk data of the pivotal materials control system, with manual follow-up based on computer summaries in the areas of direct and indirect material consumption control, material with suppliers, work-in-process, interplant accounting, purchase planning, etc."[19].

In principle, computers can be used to (i) assist in the *overall* planning, organisation, monitoring and control of production processes, and (ii) to communicate, and monitor performance at the shop-floor level. The bulk of computer applications are in the area (i), i.e. general production management and control, and not shop-floor monitoring and work-scheduling. This may be explained by a number of factors. First, small firms based partly on the use of household labour (as is the case in Prato), are likely to have only a limited demand for computer use for shop-floor monitoring which would be

18. See Shi Guangchang (1983).
19. UNIDO (1982), p. 26.

much more economical at larger scales of centralised production. Secondly, tighter and more detailed control at the shop-floor level requires adaptation of broad management systems and software packages to particular firms' specific requirements. Higher development and utilisation costs are likely to result from such adaptations.

A Case Study of the Textiles and Garments Industry

Among the many possible case studies of integration possibilities in industry we have chosen the textiles and garments industry. This, for a number of reasons.

First, this industry is important in almost all developing countries. Secondly, it is an example of a relatively "traditional" industry, i.e. one in which many of its production processes (such as pattern cutting) have not altered radically over a long period and in which traditional forms of work organisation are often found. Thirdly, data are perhaps more plentiful than in the case of other industries. Finally, the industry is export-oriented with important implications for international comparative advantage.

The advantages of applying microelectronics to the production of textiles and garments from the perspective of developed countries are shown (for each stage in the production process) in Table 2, namely, higher output, greater machine productivity, higher product quality and lower labour costs.

It is useful to consider how some of the advantages of new technology listed in the table find applications in specific developed country contexts. We shall briefly consider an example from Italy.

Prato Textiles Industry (Italy)

The example of the textile industry in the Prato region (near Florence) of Italy is particularly interesting for the potential applications of new technologies in developing countries. The Prato industry is characterised by a highly decentralised structure. It consists mainly of small-scale firms which on average employ no more than 5 to 6 workers : the bulk of them are family-based. A typical Prato weaving mill consists of a small workshop using an average of 2-3 looms operated by the owner and his family. The smallest size class (employing up to 10 workers) accounts for 91.8 per cent of establishments and 42.5 per cent of total employment in the industry (see Table 3). In absolute terms, at present, about 70,000 workers are estimated to be employed in Prato textiles. An additional 20,000 persons are engaged in support-

TABLE 2

New Technology Applications in Textiles and Garments

Process	New Technology Applications	Advantages
A. *Textiles*		
1. Fibre preparation	– automation and computer application	Increased speed of carding and drawing
2. Spinning	– open-end spinning[1] which integrates roving, spinning and winding – automatic doffing (unloading) machines	Higher output/productivity, product quality, operational speed; lower floor-space and energy requirements Economy of operators
3. Weaving	– shuttleless looms – water and air-jet looms(?)	Higher machine productivity, faster speeds, fewer auxiliary operations
4. Finishing and dyeing	– computerised continuous finishing	Lower unit labour costs and higher quality
B. *Clothing (Garments)*		
5. Cutting	– laser beams/detectors – NC-cutting devices (self-programming robotics) – water-jet cutting	Faster speeds, reduction of raw material wastes, greater flexibility
6. Design/Pattern making	– CAD devices – NC-machines	Better and more flexible lay-out designs and patterns
7. Sewing	– ultrasonic (sound waves) sewing without thread – button sewing systems – NC-sewing systems; automatic transfer lines[2]	Especially useful for synthetic materials; automatic transportation of cloth through sewing operations saves on operators and speeds up output
8. Knitting	– double-knit machines – 'Presser Foot' machine which produces partially finished garments	Increase in machine speeds and output; greater design flexibility through computer control

1. In a recent sale of open-ended rotors by Toyo Menka Kaisha (Japan) to China, it was stated that they would save 40 per cent of manpower over ring spinning. (See *Textile Asia*, June 1979).

2. Research is at present being undertaken to replace sewing altogether by textile bonding agents and meltable interlining fabrics. A recent device pioneered by Cluett Peabody in the US can pick up a piece of fabric and position it for sewing, thus eliminating the need for a sewing machine.

Source: Adapted from UNCTAD (1981).

ing services (e.g. transport, customs, banks, and independent textile mecha-
nics)[20].

A distinctive feature of the Prato textile industry is the use of rags (discarded raw materials) by recycling to produce re-generated wool. The waste
materials continue to be used although their importance has somewhat declined with the advent of new man-made fibres. Their continuance seems to be
due partly to their apparent capacity to contribute towards lower costs of
production. For example, it is reported that careful grading of the rags and
other waste material and the production of "fancy" fabrics eliminates the
need to dye the yarn : instead, original colouring can be used thus leading
to considerable economy[21]. *The Tecnotessile* (Textile Technology Research
Institute) is also engaged in a research project on the use of the soft nap
waste left over from certain textile processes.

Sub-contracting between firms, large and small, in respect of different
operations (viz. weaving, winding and carded spinning) is also quite common.
According to some estimates, the share of sub-contracting in some operations
by the woollen mills may be as high as 70 per cent of total production if
account were taken of the role of the "merchant converters" who are
generally engaged in only warping and finishing.

The sub-contracting started in the region during the economic crisis in
the fifties when redundant workers set up their own small shops, leasing
or buying old machines from larger firms. However, today the relationship
between the parent companies and the sub-contractors is based much more
on the proper division of labour and decentralisation of operations. Hardly
any firms in Prato are fully integrated in the sense that they specialise in all
operations ranging from the processing of raw materials to the stage of the
finished product.

A distinction is made in Prato between two types of sub-contractors,
viz. the "pure" ones who are set up independently of the parent firm (which
does not supply any initial capital) but works for it in carding, weaving and
finishing, etc. The second category of sub-contractors is more dependent on
the parent firm for initial establishment, supply of equipment, technical assistance etc. The latter type is created by large woollen mills with a "working
partner" who is a technically-qualified person capable of running the production unit. This unit is like a small company with several shareholders
including the "working partner" whose fortune is thus directly linked with
that of the parent company. The market for the output of the sub-contrac-

20. Mazzonis *et al.* (1983).
21. Lorenzoni (1979).

tors is fully assured by the woollen mills, thus reducing management, organisational and marketing problems.

The growing demand for the products of Prato textile industry has maintained a healthy and mutually beneficial relationship between the contracting parties. In a few cases, it has even been reported "that product diversification has been undertaken for the purpose of offering longer runs of work to sub-contractors, with the aim of securing the services of those with greater expertise"[22]. The parent companies are satisfied with the technical expertise of their sub-contractors of both artisan and non-artisan types, and with their ability to adapt to changing fashion requirements.

The sub-contracting arrangements also provide insurance against risk for small-scale firms, and stimulate them to introduce technological modernisation which may be partially financed by the parent companies. The plant modernisation (introduction of modern looms and spindles, etc.) is known to be quite impressive among small-scale firms and artisans. It is estimated that out of a total stock of 12,000 to 15,000 looms in Prato, about 1,000 have been replaced annually for the past several years.

TABLE 3

Textile Industry of Prato

| | Analysis by Size of Employment, (1975) | | | | | | | | | |
| | 0–10 | | 11–20 | | 21–50 | | 50–100 | | 101 or more | |
	% Est.	% Emp.	% Est.	% Emp.	% Est.	% Emp.	% Est.	% Emp.	% Est.	% Emp.
1) Textile	91.1	39.7	4.3	12.5	3.2	19.5	0.9	11.5	0.5	16.8
2) Knitwear	95.5	65.0	2.6	11.2	1.4	13.0	0.4	7.8	0.1	3.0
3) Total (1+2)	91.8	42.5	4.1	12.9	2.9	18.8	0.8	11.1	0.4	15.3

Est. = Establishments

Emp. = Employees

Source : Taken from Andrea Balestri, *Industrial Organisation of the Manufacturing of Fashion Goods : The Textile District of Prato (1950-1988)* – an M.A. Thesis of the University of Lancaster (UK), October, 1982.

Of the total stock of looms, more than 50 per cent are new shuttleless looms, including a few air and water-jet looms. There are over 300,000 spindles most of which are for ring spinning. Although open-end spindles (the new technology) are less common, they are being increasingly used by the larger-

22. *ibid.*

scale firms. According to the Chairman of *Tecnotessile*, his organisation has achieved a breakthrough in the application of open-end spinning technology (originally patented for cotton) to wool and wool-reclaimed fibres[23]. The initial difficulties encountered in utilising the short woollen fibres seem to have been overcome through a joint research project between *Tecnotessile* and a selected number of local firms. This innovation is considered particularly important for the Prato district which is the largest wool and reclaimed wool - processing centre in the world.

In the case of finishing and dyeing, the technique of press dyeing in small volumes has replaced traditional techniques. The processes of both dyeing and finishing are now computerised. The computer applications, however seem to be more widespread in marketing, banking, and information exchange, etc. It is now proposed to introduce a "teletext experiment" under which computers will be used with terminals in all strategic points like banks, *impannatori* (a coordinator and trade broker who keeps track of the worldwide new market for Prato textiles, and fashions, and designs), customs and transportation, etc. These terminals will supply information on raw material and product prices, fashions, new products and designs, foreign currencies, etc.[24].

The Prato textile industry produces largely for the export market. In 1981, of the total output value of 1,600 million US dollars, 1,100 million dollars worth of goods were exported to Europe, United States, Africa and Asia. The international competition to Prato textiles today puts a high premium on productivity and product quality both of which necessitate the application of new technologies.

Thanks to sub-contracting and cooperative forms of organisation, in Prato new technologies are being introduced in production notwithstanding the predominance of small firms. However, the speed in the application of these innovations is somewhat slow on account of the dispersion and smallness of firms, and their lack of knowledge about the existence and usefulness of new technologies. Although the small firms cannot *individually* afford the high cost of new machines, their cooperative organisation enables them to take advantage of these innovations. Close cooperation among producers is assured by such associations as the Industrial Union, Association of Artisans, and the Association of Traders. Furthermore, fiscal and social security legislations protect small firms which are entitled to medium-term loans at subsidised rates. It is estimated that during the last ten years, the Government

23. Private communication.
24. Mazzonis *et al.* (1983), *op. cit.*

loan funds have enabled financing of up to 50 per cent of investment in new machinery made by the weaving establishments of Prato.

In Prato, the introduction of new technology does not seem to have led to any labour displacement. In fact, in contrast to the widespread decline in textile employment in most industrialised countries since 1973, the Italian woollen textile industry (dominated by Prato) registered an increase in its employed labour force. Between 1970 and 1975, the number of textile employees increased by 12 per cent in Prato but declined by 7 per cent in France, 15 per cent in the UK, and 26 per cent in the Federal Republic of Germany[25].

The factors explaining this exceptional situation in Italy are not very clear. However, it is plausible that sub-contracting and the small-scale nature of industrial organisation ensured flexibility and adaptability of production to changing demand patterns and fashion requirements. Furthermore, the apparently successful use of discarded rags blended with reclaimed wool thus improving yarn quality, may have enabled the industry to expand production and exports at relatively low costs.

B. *Biotechnology*

The two industries in which biotechnology applications seem to be promising are : mineral industries and pharmaceuticals. Some microorganisms have a special ability to feed on inorganic and organic deposits and aid in the process of leaching and purifying mineral deposits. It is estimated "that world-wide copper recovery from ore wastes through microbial activity results in 300,000 additional tons of metal per year"[26]. Under the Cartagena Agreement, the Andean Development Corporation has installed a pilot plant in Peru to obtain 25,000 tons of copper by microbial leaching technology. It is also estimated that the two largest copper mines in Chile can attain a 7.5 - 10 per cent increase in annual production of coopper through the application of existing microbial leaching technology[27]. Thus, mineral deposits which cannot be economically exploited by conventional mining and chemical methods can be extracted by microbial treatment.

In the pharmaceutical and medical industry, the scope for such products as interferons with significant influence on cancer and viruses and insulin for treatment of diabetes, represents an important industrial potential of genetic engineering. Bio-engineering can also enable economical manufacture

25. Balestri (1982).
26. Bosecker and Kuersteu (1978).
27. Acevedo (1979).

of many drugs and vaccines on a large-scale. It is expected that many of these drugs produced with new technology would become commercialised in about five years[28].

C. Solar Energy Technologies (SET)

Much of the literature on solar energy technologies (SET) is related to their actual or potential application to the rural sectors of developing economies. However, it may just as well be possible to apply SET to urban areas of the tropics. For example, flat-plate collectors can provide hot water for cooking and washing for suburban areas and in the urban informal sector. In fact, solar water heaters are being widely used in Cyprus, Israel, Japan and the West Coast of the United States. Photovoltaic applications are also speculated for lighting of homes and refrigeration for food preservation.

One of the few concrete and well-documented examples of the applications of solar devices in developing countries is that of the solar-powered educational sets in Niger. The experiment was undertaken in 1968 with an installation of an experimental solar panel to power the television set at a school near Niamey. It is reported that by 1973 about 800 students in 22 classes were receiving instruction through solar-powered television sets. It is expected that by 1985 over 80 per cent of the population would have been covered by solar-powered educational programmes[29].

Another example of a commercialised solar technique is that of solar-heating of residential buildings in urban areas of both advanced and developing countries like Cyprus, Israel and Niger.

Despite the above examples of commercial application, solar technology of flat-plate collectors and photovoltaic cells is still relatively new and the full benefits of current R and D efforts, especially for industrial applications are yet to be realised. One potential application of solar technology in the future could be the heating of factory buildings in the developing countries.

However, few of these potential applications in developing countries are likely to become a reality so long as the costs of solar devices are high relative to incomes of the urban and the rural poor. At present, the photovoltaic systems are said to be marketed at "prices of $ 20 - $ 30 per peak watt corresponding to $ 100,000 - $ 150,000 per average kilowatt installed capacity". It is expected that these prices will decrease by as much as a factor of four within the next few years[30]. The minimum cost of a flat-plate collector is said

28. Ventura (1982).
29. National Academy of Sciences (1976a), p. 100.
30. ibid., p. 28.

to be in the range of $ 150 - $ 200 per square metre (in current US dollars); commercially installed systems may cost as much as $ 500 per square metre[31]. Until such time as the cost is substantially reduced a widespread application of solar devices in developing countries is unlikely.

3. *The Integration Matrix*

Our discussion in this third section can now be displayed in the form of a 4 × 3 matrix (Table 4) which shows the combination possibilities of each category of the emerging technologies with the four users of traditional and new technologies that we have identified. It should be stressed that the matrix is to be interpreted as suggestive of further research directions rather than as offering firm conclusions. That is, it offers an *approach* to the problem of combining traditional and emerging technologies and a research agenda rather than a set of concrete projects (more on the research implications below).

IV. *Some Limitations in the Application of New Technologies to Developing Countries*

From the point of view of developing countries, a number of limitations to the advantages of new technology applications can be listed.

a) *Product Quality Considerations*

Though quality improvement may also seem to be a facilitating factor in the adoption of the new technology in the context of developing countries, the issue is more complex because the improvements in quality that are desired in high-income countries may not be those that the majority of those living in relatively poor societies would favour.

The problem is illustrated in the oversimplified form of Figure 1. In the initial situation, denoted by point 1, only the "old" good exists and this allows consumers to obtain OA of "low-income characteristics" per dollar and OB of "high-income" characteristics. As a result of the new technology, the characteristics available per dollar will change. One case, represented by point 2, is where the new technology allows more of both characteristics per dollar than the previous technology. Here, of course, there will be no constraints on the diffusion of the new products even in low-income markets.

31. UNCTAD (1978).

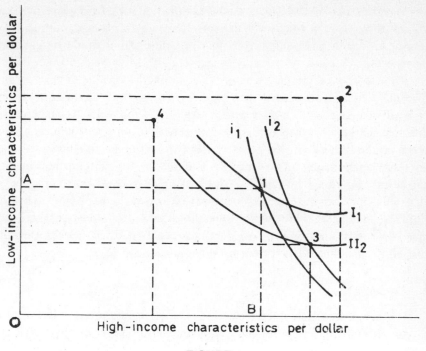

FIGURE 1

Case 3, however, shows a situation in which the new product is preferred by those in rich countries but not by the majority of consumers in the low-income societies (who prefer point 4 to point 3 because it embodies a higher proportion of low-income characteristics).

This point is well-illustrated in Mytelka's study (1981) of technology choice among textile firms in the Ivory Coast. She concludes that "to a large extent, the choice of technique in export-manufacturing is a function of international standards set for a given product — denim — in the case of COTIVO, polyester shirt fabric in the case of Gonfreville... *A far lower standard is acceptable on goods traded locally*. To meet that standard, less sophisticated production techniques are required. A wider choice of technique is thus available and the techniques tend to be more labour-intensive and less costly to purchase. The choice of more sophisticated techniques, moreover, increases the need for expatriate managers, reduces the learning effects and increases the wage bill thereby raising the costs of production"[32]. In one

32. Mytelka (1981), p. 78 (emphasis added).

interesting example cited by Mytelka (1981), an Ivorian textile firm chose printing machinery that was more closely adapted to local (and other African) markets than those in Europe. What this meant was a process which did not stabilise the cloth nor sanforise it up to the levels that would be required for exports to Europe. As a result, costs were able to be held down.

Undoubtedly, some of the improvements in the quality of textiles and garments that are associated with microelectronics will be of type 2, but there will also be many which are more like type 3 and which, in the absence of export markets to absorb this type of quality change, will impose a constraint on the adoption of the new technology in developing countries. (Sometimes even high-income *export* markets require the type of finishing quality that is only possible "by hand"). The Ivory Coast experience stands as a warning in this respect. For it was found that new technologies in the textile and wood industries, which were originally oriented to export markets, became too costly for domestic production when firms were forced through changed circumstances to reorient production to the local market.

b) *Skill Requirements*

The new microelectronics technologies in garments and textiles appear to require advanced management and maintenance skills for their efficient operation[33]. For example, introduction of the shuttleless loom in the Brazilian textile industry "raised the level of skill required to carry out maintenance tasks"[34]. And even among the UK textiles firms using microelectronics in their production processes which were part of the survey by Northcott and Rogers (1982), 17 per cent are reported to have experienced difficulties as a result of skill shortages[35].

In many cases the differential skill availabilities between rich and poor countries will mean that the latter cannot take as much advantage of the new technologies as the former. Alternatively, where the new technologies are — in spite of skill shortages — nevertheless adopted, the result will tend to be X-inefficiency i.e. the attainment of less than the maximum possible output. In some cases the new technique may even have to be abandoned.

c) *Systems Dependence*

The highly skewed distribution of the benefits of the Green Revolution

33. Hoffman and Rush (1982).
34. Acero (1983).
35. Northcott and Rogers (1982).

was due in large measure to the dependence of the HYV technology on a wide range of complementary inputs and infrastructure (see Section V below). Much the same problem may beset the new microelectronics technologies. According to Atul Wad (1982) "Microelectronics has much in common with the hybrid varieties of the Green Revolution. It is information-intensive, technologically sophisticated and though cheap in itself, is highly dependent on costly auxiliaries such as sensors, software and peripherals to be truly effective"[36]. The extent to which these forms of systems dependency will constitute a constraint on the adoption of the new technologies by developing countries is at this stage a matter of surmise. But it is suggestive that even in a developed country such as the UK quite severe problems have arisen in terms of both the availability and performance of sensors and software[37]. Indeed, "a common complaint in many potential areas of microprocessor use is that sensors and actuators that are needed to inform the microprocessor and carry out its instructions do not exist"[38].

In relation to CAD systems, Kaplinsky (1982) has noted that users require a great deal of back-up support with respect to software services from suppliers until the systems are able to function efficiently (especially when such equipment is unfamiliar to the user)."Consequently because of the established presence of CAD suppliers and systems in developed countries, it is likely that at least in the foreseeable furute the technology will spread less rapidly to developing countries than to developed ones. It comes as no surprise, therefore, to learn that of the CAD systems surveyed in the wider study only 32 out of over 6,000 systems have gone to developing countries. And of these 32 most went to either TNC subsidiaries in the petroleum industry or to governments aiming to map their countryside... to assist in counter-insurgency programmes"[39].

d) *Factor Costs and Market Size*

High capital costs have constrained the diffusion of the new technologies to smaller firms in the developed countries[40], and in the Third World, where average firm size in industry is generally much lower, this constraint will

36. Wad (1982).
37. Northcott and Rogers (1982), *op. cit.*
38. Lund (1982), p. 13.
39. Kaplinsky (1982), p. 51.
40. On the skewed pattern of adoption in the UK, see Northcott and Rogers, *op. cit.*, p. 14. For the Australian case, see Government of Australia (1980), vol. 2, p. 201.

be even more pronounced and the pattern of diffusion (and concentration) consequently even more skewed.

Moreover, among developing-country firms of an equivalent size to those in developed countries that have been able to afford the new technologies, there will almost certainly be difference in rates of diffusion on account of differences in factor costs.

As in the case of new products described in Figure 1 above, the new technologies may in some cases entirely dominate the old i.e. when they are more profitable at *all* relevant factor price ratios. But in other cases the new techniques will be profitable at the factor prices prevailing in the rich countries but not those in the poor. Considering that many of the new technologies in garments and textiles are sharply labour-saving and that the ratio of the difference in labour-costs between the rich and poor countries can be as much as 18 : 1[41], quite a few of these technologies may well fall into this category. In the past at least, this has often been true. Evidence for the weaving industry, for example, indicates that "research on improving production has not led to equipment that dictates the disregard of older methods of production in the developing countries. Indeed, the newer equipment is (privately) profitable only at wages considerably higher than market wages in the richer Latin American countries"[42].

Furthermore, some of the new technologies may be more profitable only at a scale of production that is large relative to the size of the developing country market. It appears, for example, that some of the new cutting technologies require mass production to offset their high capital costs[43].

All of the criteria considered above suggest that the new technologies will tend to find easiest application in the more affluent of the developing countries with substantial export markets, and within these countries one would expect to find a concentration among firms that are large relative to the size of the economy and which most closely approximate (in terms of organisation of work, management, etc.) medium to large-scale firms in developed countries. Applications to traditional producers will tend to be somewhat limited in scope. Alternative models of industrial organisation, however, such as that adopted in the Prato textile industry, seem to offer the potential for greatly enlarging the limited scope of new technology applications to traditional producers in developing countries.

41. For details see ILO (1980), p. 109. The differential between the United States and Korea is 10 : 1 even after allowing for efficiency difference. See Hoffman and Rush (1982), *op. cit.*

42. Pack (1982), p. 27.

43. UNCTAD (1981), p. 245.

V. *Experience of the Green Revolution Technology*

The experience of HYV as a "new" technology introduced in the rural/traditional agriculture in developing countries should offer lessons and parallel for the application of emerging technologies (like microelectronics and biotechnology) to traditional activities in general.

The promotion and use of these "new" technologies required laboratory research in international and national centres, arrangements for its diffusion to the farmers in developing countries and the provision of a package of inputs like seeds, fertilisers, pesticides, machinery, irrigation water and fuel and remunerative pricing policies.

Despite the revolutionary nature of this technology, its application in the "alien" environment of developing countries involved a number of implications and consequences which need recapitulation. These consequences or lessons could be examined as follows :

a) *Technological Maturity*

One can argue that at the time of the introduction of high-yielding seed varieties, the technology was not fully mature in the sense that its effects on the ecology and environment were not foreseen for several years. It has been discovered that high-yielding varieties are not disease-resistant unlike the local varieties. For example, Farvar[44] (1976) showed that in Iran the new wheat varieties were attacked and destroyed by a disease which did not affect the local varieties. Another ecological danger of the Green Revolution is the excessive use of water and synthetic fertilisers and insecticides.

b) *Discontinuity*

The HYVs did not emerge endogenously from traditional technologies used by the farmers in developing countries. Instead, they were specifically created in international and national scientific research centres. The laboratory research was designed to suit the environmental conditions of countries and regions in which it was to be applied. The practical experience with the use of this new technology shows that it called on "the cultivator to amend too many distinct aspects of his technology all at once, and to attempt a radical leap forward in which there is discontinuity between the existing and the new"[45]. This type of discontinuity puts a major burden on the small

44. Farvar (1976).
45. Pearse (1980), p. 180.

TABLE 4

A Matrix of Integration Possibilities

Users Emerging Technologies	Microelectronics	Biotechnology	Solar Energy Technologies
1. Traditional agriculture and rural non-farm activities (cottage industry)	– Measuring of fat content of milk	– Nitrogen-fixing (substitute for chemical fertilisers) – SCP (substitute for grain in animal feed) – tissue culture (crop yield increases)	– Photovoltaics (powering grain mill and water pump) – Photovoltaic systems (cottage industry, e.g. wood forming and metal-working)
2. Rural cooperatives	– Food storage control – Moisture control – Sprinkler control in irrigation	– recovery from agricultural wastes	
3. Small-scale industrial producers	– Mini-computers (for management, marketing) – quality control (through subcontracting relationships)	–	– Heating and lighting for factories – Flat-plate collectors (water heating)
4. Large-scale industrial firms	Self-monitoring factories; computer-aided design, manufacture and control; numerically controlled machine tools	– microorganisms (mining and extraction) – pharmaceuticals (e.g. vaccines) – recovery from industrial wastes	

farmers whose capacity to undertake the risks involved is low given the limited economic means at their disposal as noted above. This factor may partly explain why the yields attained in the laboratory experiments were not actually matched by the results at the field level.

Thus, the introduction of HYVs into traditional farming in most developing countries, represented the case in which the progressive farms under HYVs existed side by side with the traditional farms using traditional seeds and inputs. While the progressive farmers adopting HYVs became richer the economic plight of the traditional farmers remained largely unchanged.

c) *Systems Dependence*

HYV technology is highly dependent on a wide range of supporting services and infrastructure (e.g. irrigation machinery, fertilisers, pesticides) which are needed to reap their full benefits[46]. Most of these physical inputs and services originate from the urban industrial sector, thus leading to the dependence of the rural communities on the modern sector.

There is some parallel here to the situation regarding the application of new technologies like microelectronics, biotechnologies and photovoltaic cells. They are deeply rooted in the modern modes of urban industrial organisation and their maintenance and repair may also call for exogenous inputs into the traditional rural areas. Furthermore, their contribution to the removal of rural poverty presupposes supply and equitable distribution of a software package consisting of modern skills, appropriate management, and organisational structure and extension services, etc. For example, wherever adequate infrastructure (transportation network, roads, rural electrification, markets, etc.) and local research were provided, e.g. in the Indian Punjab, the farmers took advantage of the Green Revolution[47].

d) *Modes of Organisation/Production*

In order that the new technologies are easily assimilated into the prevailing cultural patterns of traditional societies, it is essential that they be in keeping with local customs and traditions. The experience with the Green Revolution shows that this new technology was biased in favour of the larger farmers and landlords who had better access to factor inputs, credit and other supporting services required. Since HYV technology is systems-depen-

46. Jéquier (1979).
47. World Bank (1982), *op. cit.*, p. 70.

dent, it would seem that it could benefit the smaller farmers more within the framework of cooperative modes of production under which better access to facilities and inputs could be assured through their sharing. Whether the utilisation of certain new technologies (e.g. small-scale satellite communications systems) can be better exploited through cooperative/collective organisation needs examination (we reached a similar conclusion in a different context above).

VI. *An Agenda for Future Research*

In the previous sections we discussed the scope and requirements for the successful integration of emerging and traditional technologies with special reference to developing countries. In doing this, we found a remarkable lack of attention in the current literature to the requirements of the traditional rural and small-scale producers. Actual applications of the three emerging technologies (micro-electronics, biotechnology and solar energy technologies) to the traditional sectors are at present very limited. While potential scope exists for their applications in the developing countries, it is unlikely to be exploited unless a conscious effort is made to determine research priorities in the light of the well-defined target groups of potential users which have been examined in this essay. This important point has been forcefully made in the context of solar technology in the following words :

"However, if this technology (photovoltaics) is to be transformed into something that can meet the special technical, economic and cultural constraints and needs of various developing countries, a deliberate and specific effort will be required; otherwise, the direct "transfer" to the remote village level of photovoltaic systems developed for integration into modern utility grids is unlikely"[48].

We also observed that the applications of biotechnology and solar technologies are much more concentrated in the agricultural/rural sector than in the modern industrial sector. The opposite is the case with micro-electronics, whose actual applications are at present largely confined to the large-scale industrial sectors of advanced and developing countries. What accounts for this diametrically opposite situation is not quite clear. It may be that the food and nutrition requirements of developing countries are rightly considered as the major priorities for research efforts in respect of biotechnology and solar technology. If this is so, why should there be very limited, if any, applications of microelectronics technology (which is much more advanced commer-

48. National Academy of Sciences (1976a), *op. cit.*, p. 29.

cially today than biotechnology) in the fields of agriculture and rural development? Yet there are speculations about their potential scope for measuring and controlling moisture, and food storage. Are there any technical limitations to the use of microprocessors in these fields in agriculture as compared to their applications in industry? Could the differences in economic organisation between the two sectors explain the differential application? This is an area for fruitful research.

Furthermore, the current literature on new technologies which we have reviewed hardly considers the possibility of combining different sets of new technologies in finding solutions to socio-economic problems. For example, is there no scope for blending the microprocessors with bio-engineering, biomass and solar technologies to bring about the technological transformation of the rural sectors of developing countries? In principle, such applications should extend the frontiers reached by the Green Revolution technology. Thus, in our view, this is another promising area of technical and economic research.

Just as there may be ways of combining the new technologies as a means of enlarging integration possibilities, so too may there be alternative forms of organising production in agriculture and industry which are capable of achieving this objective. We laid great stress, for example, on cooperatives in rural areas and subcontracting relationships in urban areas. The precise ways in which these and other forms of organising production may encourage integration of traditional and emerging technologies is a third major area of research.

Despite the plethora of studies on microelectronics and biotechnology, there is as yet no systematic methodology for determining the impact of these new technologies on employment in the advanced as well as developing countries. One reason for this gap is the relative absence of disaggregated sector and industry studies of actual applications based on primary data collection through surveys of industrial establishments. Yet such disaggregation is highly desirable particularly since the degree of new technology applications and their future implications are likely to vary between sectors, industries and processes, as well as between their domestic and export-orientation. We would therefore recommend empirical industrial case studies in both developed and developing countries with a focus on *ex-post* assessments and evaluations of new technology applications, as a fourth major area of research. Reviews of comparative country experiences and applications of traditional and emerging technologies, their identification and the selection of processes in which the latter might be economically viable, assessments of the technology applications in the light of R and D requirements, market potential, impact on employment and use of energy, local materials and skills, are some of the topics to be considered.

Since major R and D on new technologies is at present undertaken in the advanced countries for solving their economic and technical problems, the potential of these technologies for developing countries has not received the attention it deserves. Very little of such R and D activity is located within the developing countries with the result that the advanced countries will be the only sources of technology supplies in the foreseeable future.

VII. *Some Policy Implications*

We took note at the outset of this essay that the relevance of the New Technology Revolution to developing countries — the extent to which it becomes a reality for these countries — depends on how far it enables an increase in the productivity of traditional techniques. And in subsequent sections we showed that at present there are numerous constraints on the widespread integration of emerging and traditional technologies that severely limit the thoroughgoing rise in the productivity of the latter that is needed for an alleviation of mass poverty. To increase the relevance of the New Technology Revolution to developing countries (in the sense in which we have defined the term) there are two broad areas of policy.

The first is designed to alter the organisation of traditional production in ways that facilitate the adoption of emerging technologies, and the second is to adapt the emerging technologies to the needs of traditional producers. In this concluding section we discuss the problems associated with specific policies in each of these areas.

As we have pointed out above, the re-organisation of traditional production will need to be based on co-operatives in rural areas and subcontracting relationships in the industrial sector. Let us deal first with the requirements for the successful operation of rural co-operatives.

Probably the main problem that has to be confronted here is of a socio-political kind, namely, that "in situation of asset disparities, emphasis on rural self-help and grassroot participation — frequently leads to the control of local organisations by the rural political élite, resulting in a disproportionate allocation of the scarce development resources in their favour"[49]. The failure of the Ujamaa village scheme in Tanzania and the *Panchayats* in India can be explained, at least partly, in these terms. And the more unequal is the ownership of land, the more likely it is that rural co-operatives will be dominated or

---

49. Lele (1981), p. 56.

subverted by the élite groups. Consequently, if traditional producers are to be effectively served through co-operative organisations in such situations, a redistribution of land and political power in their favour would seem to be an essential pre-requisite. It is surely significant, as Lipton (1981) has pointed out, that "the popular success stories of rural credit — Taiwan and South Korea — followed radical equalisation of rights in land through distributive reform"[50].

If, therefore, it is possible to argue that such reform is likely to constitute a necessary condition for rural co-operatives to reach the poor, it is decidedly *not* a sufficient condition. For what is also essential is the adequate provision to these institutions of inputs, management, know-how, infrastructure, etc. These kinds of factors, according to Lele (1981), "receive relatively little emphasis in co-operative development, in comparison with a rather amorphous and mis-directed concept of existing grassroot organisations and of their potential for promoting participation and democracy"[51].

With respect to the industrial sector, the problems of altering the way production is, for the most part, now organised seem to be no less formidable. For although subcontracting relationships were an important feature of the development of many countries that industrialised in the previous century (and more recently, also in Japan and Korea), there is, in most developing countries "a tendency towards vertical integration rather than a reliance on either subcontracting or market relationships with suppliers"[52]. While the nature of the reasons for this divergent pattern of industrial development are not altogether clear, it does seem that an alteration of this tendency, in the direction of providing greater linkages between large and small firms, will require a policy of active encouragement of the latter. It will be recalled from our discussion of the Prato Textile Industry in Italy (where subcontracting is fairly common) that considerable protection and assistance is granted to the many small firms in the industry. Similar encouragement (in terms of access to credit, foreign exchange, technical assistance, etc.) will need to be given to small-scale enterprises in developing countries; at the same time, the existing discriminations against these firms should be eliminated. The Prato experience also underlines the importance of providing a package of centralised services and utilities. Herein lies a role for the government. It is unlikely that the new and emerging technologies could be effectively used in developing countries without some provision of government support and/or

50. Lipton (1981), p. 208.
51. Lele (1981), *op. cit.*, p. 69.
52. Rao (1974), p. 145.

infrastructural facilities like the teletext network proposed for Prato.

So far we have described some of the problems involved in altering the organisation of traditional production so as to facilitate the adoption of the emerging technologies. We now discuss how, instead, these technologies may be adapted to fit in with the requirements of traditional producers. This is the second broad area of policy noted above.

There are several good reasons for the developing countries to undertake the research required for adaptation of the emerging technologies themselves (though, of course, one does not wish to exclude some form of supportive international role)[53]. First, the learning effects of the research could be captured within the developing countries. The second reason has to do with our insistence on the need for research to be closely related to the needs of the target groups of potential users. Though this objective will not *necessarily* be achieved merely by the location of research in developing countries, it seems to us more likely since the necessary close links with traditional producers will then be easier to forge and maintain.

One factor that may frustrate the undertaking of this type of research in developing countries is known as the "assurance problem". That is, because of the heavy costs and substantial risks involved, each *individual* country may be reluctant to undertake the research without an assurance that others will do likewise. And since the market mechanism does not normally produce this kind of assurance, the result will be a collectively suboptimal amount of research expenditure. The solution to this type of market failure is some form of co-operative arrangement, that in this context may perhaps most effectively be sought on a regional basis. Alternatively, the research could be coordinated through the agency of a single international institution (such as UNIDO's proposed International Industrial Technology Institute)[54].

REFERENCES

ACERO, L. (1983), *Impact of Technical Change on Skill Requirements in Traditional Industries : The Case of Textiles*, SPRU, University of Sussex, (mimeo.)

ACEVEDO, F. (1979), *Applicacion de la Ingenieria Bioquimica à la Lixiviacion de Sulfuros*, Paper presented at the Seminario Internacional Sobre Procesos Especiales de la Metalurgia Extractira del Cobre, Trujillo, Peru.

53. Stewart (1981).
54. UNIDO (1979), Chapter 7.

AUSTRALIA GOVERNMENT, (1980), *Technological Change in Australia*, Report of the Committee of Enquiry into Technologieal Change in Australia.

BALESTRI, A. (1982), *Industrial Organisation in the Manufacturing of Fashion Goods : The Textile District of Prato* (1950-1980), M.A. Thesis of the University of Lancaster, UK.

BESSANT, J. *et al.* (1981), "Microelectronics in Manufacturing Industry" in Tom Forester (ed.), *The Microelectronic Revolution — The Complete Guide to New Technology and its Impact on Society*, MIT Press, Cambridge, Mass.

BHALLA, A.S. and JAMES, J. (1983) "An Approach Towards Integration of Emerging and Traditional Technologies", in E.U. von Weiszäcker, M.S. Swaminathan and Aklilu Lemma (eds.), *New Frontiers in Technology Application — Integration of Emerging and Traditional Technologies*, Tycooly International Publishing Ltd.

BIFANO, W.J. *et al.* (1979), "A Photovoltaic Power System in the Remote African Village of Tangaye, Upper Volta", *NASA Technical Memorandum*, No. 79318.

BOSECKER, K. and KUERSTEU, M. (1978), "Recovery of Metallic Raw Materials by Microbial Leaching", *Process Biochemical*, vol. 13, No. 2.

FARVAR, M.T. (1976), "The Interaction of Ecological and Social Systems : Local Outer Limits in Development" in W.H. Matthews (ed.), *Outer Limits and Human Needs*, Almquist and Wicksell.

HÉDEN, C.G. (1979), "Microbiological Science for Development: A Global Technological Opportunity", in Jairam Ramesh and Charles Weiss, Jr. (eds.), *Mobilising Technology for World Development*, Praeger.

HOFFMAN, K. and RUSH, H. (1982), "Microelectronics and the Garment Industry : Not Yet a Perfect Fit" *IDS Bulletin*.

INTERNATIONAL LABOUR OFFICE, (1980), *Report of the Second Tripartite Meeting for the Clothing Industry*.

JÉQUIER, N. (1979), "Appropriate Technology: Some Criteria" in A.S. Bhalla (ed.), *Towards Global Action for Appropriate Technology*, Pergamon.

KAPLINSKY, R. (1982), "Is there a Skill Constraint in the Diffusion of Microelectronics?" *IDS Bulletin*.

KING, K.J. (1974), "Kenya's Informal Machine-makers — A Study of Small-scale Industry in Kenya's Emergent Artisan Society", *World Development*, April-May.

LELE, U. (1981), "Cooperatives and the Poor: A Comparative Perspective", *World Development*, vol. 9 (1), January.

LIPTON, M. (1981), "Agricultural Finance and Rural Credit in Poor Countries", in P. Streeten and R. Jolly (eds.), *Recent Issues in World Development*, Pergamon.

LORENZONI, G. (1979), *A Policy of Innovation in the Small and Medium-sized Firms – An Analysis of Change in the Prato Wool Industry*, translation from Italian (mimeo).

LUND, R. T. (1982), *Microprocessor Applications and Industrial Development*, UNIDO ID/W.G. 372/14, 9 August.

MAZZONIS, D., COLOMBO, U. and LANZAVECCHIA G. (1983), "Cooperative Organisation and Constant Modernisation of the Textile Industry at Prato (Italy)" in E.U. von Weizsäcker, M.S. Swaminathan and Aklilu Lemma (eds.), *New Frontiers in Technology Application - Integration of Emerging and Traditional Technologies*, Tycooly International Publishing Ltd.

MYTELKA, L. (1981), "Direct Foreign Investment and Technological Choice in the Ivorian Textile and Wool Industries", *Vierteljahresberichte*, No. 83.

NARANG, S.A. (1981), *Genetic Engineering : The Technology and its Applications*, Report for Consultations on the Implications of Advances in Genetic Engineering for Developing Countries, UNIDO, (mimeo).

NATIONAL ACADEMY OF SCIENCES, (1976), *Technology Assessment Activities in the Industrial, Academic and Government Communities*, Washington D.C.

—— (1976a), *Energy for Rural Development : Renewable Resources and Alternative Technologies for Developing Countries*, Washington D.C.

—— (1982), *Diffusion of Biomass Energy Technologies in Developing Countries*, National Academy Press.

NORTHCOTT, J. and ROGERS, P. (1982), *Microelectronics in Industry : What is Happening to Britain?*, Policy Studies Institute.

PACK, H. (1982), "Aggregate Implications of Factor Substitution in Industrial Processes", *Journal of Development Economics*, vol. 11.

PEARSE, A. (1980), *Seeds of Plenty, Seeds of Want : Social and Economic Implications of the Green Revolution*, UNRISD and Clarendon Press.

RADNOR, M. (1982), *Prospects of Microelectronics Application in Process and Product Development in Developing Countries*, UNIDO/ECLA.

—— and WAD, A. (1981), *A Programme for Microprocessor Capability Development for African Nations – A Proposal Outline*, Northwestern University.

RAO, D.C. (1974), "Urban Target Groups" in H. Chenery *et al.*, *Redistribution with Growth*, Oxford University Press.

SHI, G. (1983), "Chinese Experiences with Applying New Technologies to Traditional Sectors" in E.U. von Weizsäcker, M.S. Swaminathan and

Aklilu Lemma (eds.), *New Frontiers in Technology Application — Integration of Emerging and Traditional Technologies*, Tycooly International Publishing Ltd.

STEWART, F. (1981), "International Technology Transfer : Issues and Policy Options" in P. Streeten and R. Jolly (eds.), *Recent Issues in World Development*, Pergamon.

STREETEN, P. and JOLLY, R. (1981), *Recent Issues in World Development*, Pergamon.

UNCTAD (1978), *Energy Supplies for Developing Countries : Issues in Transfer and Development of Technology*, TD/B/C. 6/31.

—— (1981), *Fibres and Textiles : Dimensions of Corporate Marketing Structure*, TD/B/C. 1/219.

UNIDO (1979), *Industry 2000 : New Perspectives*, New York.

—— (1982), *Microelectronics Monitor*, No 3.

VENTURA, A. (1982), "Biotechnologies and their Implications for Third World Development", *Technology in Society*, vol, 4, No. 2.

WAD, A. (1982), "Microelectronics : Implications and Strategies for the Third World", *Third World Quarterly*, vol. 4, No. 4.

WATANABE, S. (1983), *Technology, Marketing and Industrialisation : Linkages Between Large and Small Enterprises*, Macmillan.

WEIZSÄCKER, E.U. VON, SWAMINATHAN, M.S. and LEMMA, A. (eds.) (1983), *New Frontiers in Technology Application — Integration of Emerging and Traditional Technologies*, Tycooly International Publishing Ltd.

WORLD BANK (1981), "Evaluation of Microbial Technologies Involved in Fuel Production, Agriculture and Forestry", *Science and Technology Report Series*.

—— (1982), *World Development Report for 1982*, Oxford University Press.

CHAPTER SEVEN

PUBLIC POLICY IN AN ECONOMY WITH DIFFERENT TYPES OF AGENTS

By J. L. Enos*

Introduction

Economists have long puzzled over the relationship between individual economic agents, on the one hand, and the collectivity, on the other. Systematic investigations of the relationship have followed one of three approaches; the classical, the bureaucratic and the game-theoretic. The first, pioneered by Adam Smith and Walras, perfected by their protegés, and, in recent times extended by those revelling in rational expectations, revealed the macro-economic properties of the atomistic economy. The second, espoused by Lange and Lerner and quantified by economists in both Western and Eastern blocs, has resulted in the planning algorithms of Dantzig-Wolf and Kornai-Liptak. The third, commenced by von Neumann, has culminated in the convergence theorem of Debreu and Scarf, which exposes, in deductive theory at least, the ubiquity of the perfectly competitive general equilibrium.

All three approaches to the theoretical study of an economy comprising individual agents have succeeded by imposing uniformity on their behaviour. Any action other than profit-seeking in precluded; any intervention with the aim of changing the "rules of the game" is proscribed. To be sure, theorists must be prepared to abstract from reality if they are to devise models which yield interesting deductions, but one may well ask if such a grand abstraction from the real variety of economic types is necessary. Is it possible to contemplate the overall performance of an economic system comprising agents with differing aptitudes, engaging in differing activities? Is it possible, next, to deduce how such an economic system would respond to the interventions of a government? Is it possible, still further, to develop a set of policies which could guide a government determined to influence the structure of the economy? These are questions which we will attempt to answer in the following essay.

The context within which we will address the above questions is that of a developing country whose government faces the task of allocating most

* Magdalen College, University of Oxford.

The author wishes to thank Peter Hall, Sharon Rochford and Jonathan Baldry for their comments, while absolving them of responsibility for any remaining errors or lack of clarity : the author is solely responsible for these.

efficiently the country's scarce technical and administrative resources among two competing activities — absorbing foreign modes of production, chiefly technological, and creating indigenous modes. To put it more simply, the alternatives are to imitate or to innovate. Both activities are carried out by self-interested firms operating in an environment whose rewards are influenced by government, and possibly also by public-sector firms. The organization of the essay is as follows : in the next section we will discuss the phenomenon of innovation, as it manifests itself in developing countries. After that, a section will be devoted to the formulation of a game-theoretic model representing the allocation of the country's technological resources among the different types of firms. Solutions to the model, with and without government intervention, will fill the penultimate section; and in the final section the implications will be drawn for public policy and for further research.

The Nature of Innovation in a Developing Country

Economists studying innovation have tended to focus on the developed countries, for the obvious reason that it is in the developed countries of Western Europe and North America that most modern industrial techniques and products have first appeared. Moreover, since manufacturing techniques move fairly freely from one developed country to another, the developed countries as a whole have been considered to be the relevant universe within which innovations arise. Consequently, innovations have generally been defined in terms of the first commercial production and distribution, anywhere in the world, of a novel product; or the first commercial application, again anywhere in the world, of a novel process or form of organization; or the first exploitation of a novel market.

Under such a definition, innovation has hardly ever occured, and will hardly ever occur, in a developing country. In its conventional sense, therefore, the term innovation is irrelevant for our purposes. In this situation, we are left with three alternatives : we can redefine the term to suit our purposes; we can provide a qualifying adjective to the term; or we can invent a new term (which would be an innovation in itself, according to the conventional usage of the term). The choice is somewhat arbitrary, but will be postponed until we have considered what takes place in a developing country that is to be the site for the production and distribution of a formerly-novel product, or for the adoption of a formerly-novel process, etc; in conventional terms, what takes place in a developing country that intends to imitate.

There are two polar cases of imitation, so far as the developing country is concerned. At one extreme, the imitating country can enlist the services of the foreign organization that introduced to the world the novel product, etc; or, for the developing country an equivalent choice, another foreign organization that has, in between the date of the original innovation and the present, succeeded in imitating the manufacture of the novel product, etc. The foreign organization then undertakes, entirely on its own, on behalf óf and in the developing country, all the activities necessary to produce and distribute the product, etc. At the other extreme, the developing country can mobilize its resources − its own scientists, its own engineers, its own financiers, etc. − and replicate the entire process undergone by the foreign innovator or imitator.

Between these two extremes lies a spectrum, along which developing countries tend in practice to fall. Developing countries with few domestic resources usually end up towards the first extreme of complete dependency; those with abundant domestic resources usually end up towards the second extreme of independence. Developing countries adopting complex and sophisticated technologies usually end up towards the first extreme; those adopting simple technologies usually end up towards the second. Developing countries whose leaders wish to maintain close and profitable relationships with one or more developed countries usually end up towards the first extreme; those whose leaders wish to be self-sufficient usually end up towards the second.

The reasons underlying the choice of the point along the dependency-independency spectrum are many and beyond the scope of this essay; they have been studied for a set of sophisticated manufacturing processes in one developing country by this author and his colleague (see Enos and Park, forthcoming). What is germane to this essay is what we should call the collection of activities that are thereafter undertaken in the developing country, and whether or not what we call them will vary depending upon where along the spectrum the developing country chooses to settle.

Let us commence by looking at the polar case of independence, in which the entire process of innovation is replicated in the developing country, from the recognition that an invention has commercial potential to operation on a relatively large scale. Such a case, which we could call local innovation (meaning innovation within a limited geographical area), would arise very seldom because it would require an ignorance of the original innovation, which was, by definition, a commercial success. We will neglect any such case of reinvention and re-innovation.

A point on the spectrum, close to the extreme of local innovation, worthy of consideration is what might be called pseudo-innovation. Pseudo-innovation differs from innovation proper only in that the agent carrying it out knows that the novel product/process/market/form of organization has already been exploited elsewhere. Otherwise, the pseudo-innovator is required to undertake all the activities that the proper innovator did : assemble the resources, conduct the research and development, evaluate the market, file for patents, carry out a pilot operation, etc. His roles are similar to those carried out by the entrepreneur in a developed country who sets out to invent around the proper innovator's patents, although the latter will probably have an easier task since the necessary resources are more readily available. Usually the pseudo-innovator in the developing country will be filling the majority of the roles for the first time : he will be assembling the resources for the first time, administering the research and development for the first time, etc. His task is likely to be at least as monumental as was the proper innovator's[1].

The pseudo-innovator will be one of our two types of economic agents. The other type of economic agent in our game-theoretic model will be the less ambitious "absorber", i.e., the firm which adopts an unfamiliar product/process/market/form of organization relying upon an experienced foreign supplier to carry out all the novel activities. Little by little the absorber may then substitute its own resources for those of the foreigner, learning in sequence how to operate facilities, how to train staff, how to design and possibly how to construct pieces of equipment, how to secure modest improvements and the existing scheme, and finally how to carry out the overall design and construction of new facilities incorporating the same state of the art.

On the innovation spectrum the absorber will lie somewhere around the middle; he does not attempt to go it on his own, as does the pseudo-innovator, but is content to let the foreign supplier carry out all the tasks initially. However, unlike the firm which remains ever-dependent upon the foreign supplier, the absorber does ultimately displace the expatriate. In the country, South Korea, whose adoption of sophisticated techniques this author has studied, most of the firms have been absorbers. The numbers of pseudo-innovators and ever-dependent firms have been very, very few, the Korean Government having preferred to allocate its scarce technical and managerial resources to the task of absorbing modern, western technology. The time may have arrived, however, when pseudo-innovators could emerge : whether or not their emergence should be encouraged is one of the issues to which our model will be addressed.

1. Even in a developed country, where resources are readily available, the imitator's costs are likely to be a substantial proportion of the innovator's (Mansfield, el. al., 1981).

A Game-Theoretic Model of a Country's Economic Structure

In order to keep our exposition simple, we are restricting the types of economic agents in the model to two : pseudo-innovators and absorbers. Let the pseudo-innovators be labelled 1 and the absorbers, 2. (The nomenclature, summarized in Table A, is drawn from Karlin (1959)). Each firm within the economy is faced with an irrevocable choice : to be a pseudo-innovator (strategy x_1) or to be an absorber (strategy x_2). In equilibrium, a proportion, P, of all the firms in the economy will have chosen to be of type 1 and the residual proportion, $(1 - P)$, to be of type 2. The economic structure of the country is then defined by the proportion P, representing the fraction of all firms which are pseudo-innovators.

TABLE A

X_i	= set of strategies available to firm i;
	$\quad$ i = 1, 2 (in model)
	$\quad$ i = 1, 2,..., n (in general)
x_i	= choice of type of firm, $x_i \, \varepsilon X_i$
P	= the structure of the economy, defined as the proportion of all firms which are of different types (in model, the proportion which are of type 1; $0 \leqslant P \leqslant 1$)
$a_{i,j}(P)$	= average return (pay-off) to firm of type i from encounter with firm of type j
$K(P)$	= pay-off kernel function
$E(x_i)$	= excepted return to the firm of adopting strategy i
N	= number of firms in economy (a continuum, whose total is normalized to unity)
π	= returns to all firms
x_0	= optimal strategy for firm of type X_1
y_0	= optimal strategy for firm(s) of other type(s)
P^0	= P equal to zero (in model); $\dfrac{d\pi(P^0)}{dP} < -1$
P^-	= P at which returns to all firms are (locally) minimized ; $P^0 < P^- < P^*$, $\dfrac{d\pi(P^-)}{dP} = -1$
P^*	= P at which returns are maximized; $\dfrac{d\pi(P^*)}{dP} = -1$
P^g	= P desired by government (in model)
P^1	= P equal to unity; $\dfrac{d\pi(P^1)}{dP} < -1$

The firms are guided in their choice by the expected returns from selling their products in competition with other firms. Competition is assumed to be of a pair-wise nature, the ith firm, say, being in competition with the jth for any particular sale. The jth firm, with which it is in competition, is assumed to emerge at random from the universe of all firms in the economy; the probability that the jth firm will be of type 1 will be equal to P, and of type 2, $(1 - P)$. The performance of the ith firm will be measured by the average return from a succession of encounters with its rivals.

The possible outcomes for firm i from any encounter are summarized in the pay-off matrix in Table B. The alternative choices of the ith firm, on which we are focusing, are tabulated on the left of the matrix; the jth firm, drawn from the universe of all other firms, along the top. The numbers in the cells of the pay-off matrix, four in our simplified case, represent the average returns to the ith firm should it choose to follow the strategy listed on the left and should its rival of the moment have chosen the strategy indicated by the column above. For example, should the i^{il} firm choose to be a pseudo-innovator and should it encounter a rival of the opposite type, its average return from the action would be $a_{1,2}(P)$, the number indicated in the upper right-hand cell of the pay-off matrix.

TABLE B

Matrix of Pay-Offs in the Case with Two Types of Firms

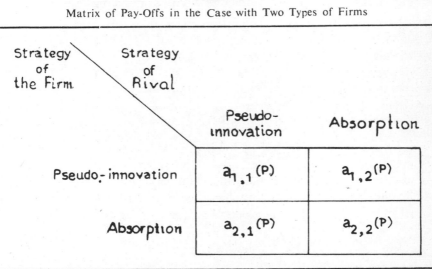

Strategy of the Firm \ Strategy of Rival	Pseudo-innovation	Absorption
Pseudo-innovation	$a_{1,1}(P)$	$a_{1,2}(P)$
Absorption	$a_{2,1}(P)$	$a_{2,2}(P)$

For the ith firm, the outcome of an encounter on the average is seen to depend upon only three factors — its choice, the choice of its momentary rival, and the structure of the economy. Its choice or strategy, and that of its rival, are indicated respectively by the row and column headings of the pay-off matrix; the structure of the economy is indicated by the variable P, of which the $a_{i,j}$ are a continuous, real-valued function over its domain $(0 \leqslant P \leqslant 1)$. The function $K(P)$, whose elements are the $a_{i,j}(P)$, is thus a vector-valued function whose domain is the unit interval, and whose range is a bounded set of real numbers.

The function $K(P)$, which translates strategic choice, at any given economic structure, into pay-offs, is called the kernel function or the outcome function : we will use the former term. In specifying a kernel function we are abstracting from cost and market conditions other than those of type of firm and pair-wise encounter. It matters not to the analysis how the firm behaves, nor how the industry is organized ,since these aspects of the economy are subsumed in the specification : all that it is necessary to state is that when such and such a type of firm encounters a firm of such and such a type within an economy of such and such a structure the outcomes, on the average, for each firm are so many monetary units of reward.

The kernel function is complete, in that it admits an outcome to every type of encounter. Moreover, the $a_{i,j}(P)$ of which it is constructed are single-valued, in that given the types of rivals and the structure of the economy, only one outcome can eventuate to the ith firm.

Applications, utilizing the kernel function have been made by geneticists studying the distribution of gene types (Maynard Smith (1982)); but since it admits of no variation in behaviour, such as coalition formation, the kernel function has not been widely used by economists (Shubik (1982), p. 382)[2]. To stay on familiar ground, it will be worthwhile to express the relations between the $K(P)$ and conventional functions. In our environment, comprising pseudo-innovators and absorbers, the expected returns, on the average, to the ith firm, should it choose to be a pseudo-innovator itself, would be

$$E(x_1) = P\, a_{1,1}(P) + (1 - P)\, a_{1,2}(P) \tag{1a}$$

where it is assumed that the outcome from each encounter is independent of the outcomes of previous encounters. Should the ith firm choose to adopt the strategy of absorption, its expected returns, on the average, would be

$$E(x_2) = P\, a_{2,1}(P) + (1 - P)\, a_{2,2}(P) \tag{1b}$$

Before we can aggregate across firms, so as to be able to describe the simultaneous choice for all of them, it will be necessary to augment the pay-

2. The only application known to the author is that of Sharon Rochford (forthcoming).

off matrix and to specify a utility function. Considering the pay-off matrix, that of Table B reveals, in the entries in the cells, the returns to the player on the left. We shall assume that the returns to the jth firm, of the same type as the ith firm but drawn at random from the universe of all firms, will be the same as if it were the ith firm; i.e. that the $a_{j,i}$ are equal to the $a_{i,j}$. For example, if the ith firm chooses the strategy of pseudo-innovation and meets another pseudo-innovator, its rival's return from the competitive encounter will be the same as the ith's, namely $a_{1,1}$. Thus it does not matter for the outcome whether the firm appears in the left-hand side or the top of the matrix in Table B. When players can be interchanged without affecting the pay-offs, the matrix is said to be symmetric; and one set of numbers, the $a_{i,j}$, reports the returns to every player. Implicit in equations (1a) and (1b), whose left-hand sides are written without subscripts, is a symmetric pay-off matrix.

The assumption of equal return to all players of a single type, while simplifying the analysis, does violation to reality, particularly in developing countries. Real firms with the same specialization are not interchangable, receiving the same rewards for the same actions, but are almost unique as a consequence of differences in size, experience and location, and in their owners' wealth, position and connections. Representing in game theoretic format a universe of unique firms would require a separate pay-off matrix, assymmetric in nature, for each pair of contestants. Analytically, such a game might well be intractable, but even if it were tractable the requirements for data would be immense. In this essay we shall content ourselves with symmetric pay-off matrices, admitting that such simplicity is unrealistic.

Considering the utility function, we shall assume that firms' preferences are in accord with the axioms of von Neumann and Morgenstern[3]. Technically, these functions are invariant up to an affine transformation; practically, they are linear in increments of money (the pay-offs) for any single firm and comparable (transferable) across firms. Thus, a pay-off of two units is twice as valuable to any firm as a pay-off of one unit, and a pay-off of two units to one firm is worth as much to that firm as would be the same pay-off of two units to another firm.

One effect of assuming von Neumann-Morgenstern utility functions for the firms is that equations (1a) and (1b), which were written to express expectations of pay-offs, become the firm's valuation of alternative strategies. Within our simple example, the measure, to the ith firm, of the attractiveness of choosing to be of type 1 is $E(x_1)$; of type 2, $E(x_2)$. If $E(x_1)$ exceeds $E(x_2)$, then the choice of x_1 is preferable to the choice of x_2.

3. If firms' rewards are adequately measured by profits, these alone can be maximized, and the von Neumann-Morgenstern axioms dispensed with.

Another effect is that we can, by proper re-scaling, add up the pay-offs of different firms, thereby aggregating utilities. If there are N firms in the economy, a fraction P of which choose to be pseudo-innovators and $(1 - P)$ absorbers, the total returns to all the firms will be the summation $\Sigma N[PE(x_1) + (1 - P)E(x_2)]$. Designating total returns by the letter π, we define them as :

$$\pi \equiv \Sigma N[PE(x_1) + (1 - P)E(x_2)] = \pi(P) \qquad (2)$$

the term on the far right reminding us that returns are a function of the structure of the economy.

For analytic purposes we have made assumptions which enable us to accommodate the problem of the firm's choice within a class of infinite games with continuous pay-off kernels played over the unit interval. We have already described the competitive environment as involving, for each firm, repeated encounters with different rivals drawn randomly from the universe of all firms, P of which are pseudo-innovators and $(1 - P)$ of which are absorbers. The pay-off matrix, whose elements are units of transferable utility, is the same for each encounter, the specific pay-offs depending upon the ith firm's strategy, its rival's, and P. These conditions are sufficient to assure a solution (Karlin, (1959), Section 2.2).

The general nature of the solution can best be imagined if we concentrate on the alternative choices (strategies) available to the ith type of firm, given the distribution of firms that it will encounter. If this distribution is optimal for all the j types of firms ($j = 1, 2, \ldots, i + 1, \ldots, N$), it can be represented by an optimal strategy y_0. In response to its rivals optimal strategy of y_0 the best strategy for the ith type of firm is x_0, fulfilling the condition :

$$K(x, y_0) \leqslant K(x_0, y_0) \qquad (3a)$$

where x is all possible strategies. In words, $K(x_0, y_0)$ will yield firm i at least as high an expected return, over all contests, as any other alternative, when its rivals are in their turn, choosing their best alternatives.

In our model with a symmetric pay-off matrix, rivals who happen to have chosen the same alternative as the ith firm will receive an identical pay-off, i.e. $K(x,y) = -K(y,x)$. In the terminology of games theory, the rival is assumed to aim at minimizing its maximum loss, which means choosing its optimum strategy, y_0, in the face of the ith firm's choice of x_0, such that :

$$K(x_0, y_0) \leqslant K(x_0, y) \qquad (3b)$$

for all possible strategies y. Combining the inequalities (3a) and (3b), we obtain the relation describing the best choices for both the ith and the jth firms from among all their alternatives :

$$K(x, y_0) \leqslant K(x_0, y_0) \leqslant K(x_0, y) \qquad (3c)$$

Given our presumption, stated earlier, that each firm's objective is to maximize the expected returns, and that each chooses so as to achieve its objectives; and given our convention of thinking of the ith firm's returns as positive and its rival's as negative, we discover that the optimal strategy pair $x_0.y_0$ conforms to the equation :

$$\max_{x \varepsilon X} \min_{y \varepsilon Y} K(x_0, y_0) = \min_{y \varepsilon Y} \max_{x \varepsilon X} K(x_0, y_0) \qquad (4)$$

This is the idea of the Nash equilibrium, that the ith firm is making its best choice, given the best choices of all the other firms. All are optimizing simultaneously.

We also discover that the optimal strategies coincide (*ibid.*, p. 28). The optimal strategy for the ith type of firm, confronting a succession of rivals distributing themselves according to y_0, will be to choose the strategy $x_0 = y_0$. In our model the strategy y_0 is equivalent to a distribution of firms within the economy equal to P_0, chosen from among all possible distributions P. When all its rivals are distributed according to P_0, the ith firm will find that whichever choice it makes among its alternatives its expected returns will be identical. From the overall point of view, the ith firm's choice is immaterial, since all the other (many) firms have distributed themselves optimally among the possible alternatives.

Even though the existence of solutions to infinite games with continuous pay-off kernels defined on the unit interval can be assured, the solutions are by no means easy to obtain. Game theorists find it useful to restrict the nature of the pay-off function so as to be able to examine the particular solutions; one convenient restriction is to the class of convex continuous kernels. The convexity can be visualised by focusing on the pay-offs associated with the rival's alternative choices. In the format of the game, the rival is seen as attempting to minimize the return to the ith type of firm. In making its choice, the rival casts its eye over the entire kernel, with the intention of minimizing $K(x, y)$ for the ith firm's every x. If the function $K(x, y)$ is strictly convex in y for each x in its domain there will be a unique optimal strategy y_0 for the rival. Reciprocally for the ith type of firm which is attempting to maximize its minimum return, if the pay-off kernel is strictly concave in x for each y of its rival, it will have a unique optimal strategy x_0.

Economists are like game theorists in preferring systems with unique solutions. Their analogue of the games theorist who restricts himself to games with convex, continuous kernels is the international trade theorist who restricts himself to strictly concave product transformation schedules. This is equivalent to assuming the function $\pi(P)$ in equation (2) is strictly concave. If one traces the function $\pi(P)$ back to its source in equations (1a) and (1b) one finds that strict concavity in $\pi(P)$ is consistent with strict concavity in the pay-off kernel for x with respect to each y.

But can we be content with strict concavity for x with respect to each y, and strict convexity for y with respect to each x? In other words, can we be content with an assumption of a strictly convex pay-off kernel function? Are we justified in simplifying the reality of an underdeveloped economy so as to be assured of obtaining a unique solution to our model? Our arguments in the next section will be that we are not justified in making such a simplification and that such a simplification removes from consideration important policy options available to the governments of developing countries.

The Firm's Choice in a Model with a Generalized Pay-Off Function

Our model of the choice confronting a firm indicates that the outcome is dependent upon the alternatives open to it, the alternatives open to its rivals, and the overall structure of the economy. What we now wish to do is to imagine how the expected returns to the firm might vary as the structure of the economy changes, and then to see how such changes would affect the firm's optimal strategy. In terms of games theory, we will reject the assumption of convexity in the pay-off kernel function, while retaining the prior restrictions of continuity over the unit interval.

How would we expect the returns to individual firms to vary with varying economic structure? Let us address this question first within the narrow context of innovative structure and subsequently within the broader context of a generalized structure, i.e. a structure defined in multiple dimensions. Let us also consider it within a static environment.

As an example from the spectrum of developing countries we might take one of the relatively large and increasingly sophisticated nations with a substantial industrial sector producing goods which are internationally competitive : i.e. one of the newly-industrialized countries. Such a country is likely to have a stock of well-educated scientists, engineers and managers, to be employing already modern techniques imported from abroad, to be expending considerable amounts of foreign exchange on licenses, royalties and other forms of technical assistance, and to be contemplating the encouragement of domestic research and development. What shape would the total returns function $\pi(P)$ for a newly-industrializing country take on?

Starting at the boundary representing the economy without innovative firms (where $P = 0$), what returns would we expect to accrue to the first potential pseudo-innovator? In such a state, the potential pseudo-innovator would encounter a steady succession of absorbers, with each of whom it would be in contention. Its unit costs, should it choose to be a pseudo-innovator, would undoubtedly be greater than those of its rivals, absorbers, because it would be maintaining the same production and distribution facilities, and a research and development department as well. Assuming that some outside agency did no more than meet the research and development costs or that the research and development did not generate a new product or process which would confer some substantial competitive advantage, the returns to pseudo-innovation would be less than those to absorption. In terms of the pay-off matrix of Table B, $a_{1,2}(P^0)$ would be less than $a_{2,2}(P^0)$. The optimal choice for the rational firm would be to be an absorber; i.e. the strategy x_2 would be optimal for the ith firm in a universe in which all other firms would choose the optimal strategy x_2. The boundary point P^0 would be a stable solution from which no firm would voluntarily depart. Any departure, towards pseudo-innovation, by a single firm, would reduce its returns, and the total returns for all firms as well.

This part of the argument rests upon the assumption that there are not sufficient external economies, arising from the research and development carried out by the first pseudo-innovator, to raise the returns to all firms in the economy consisting of one pseudo-innovator and $(N - 1)$ absorbers to a point exceeding the total to an economy of N absorbers. In our symbols, this is equivalent to assuming $\pi(P^0) > \pi(P^{0+})$.

In an economy just beginning to develop, in which the resources needed for innovation are very scarce, total returns would in all likelihood be maximized at the boundary where all firms are absorbers; i.e. $\pi(P^0) > \pi(P)$, $0 < P \le 1$. In an economy which is newly industrialized, however, there may be some point, say P^*, $(0 < P^* < 1)$, where total returns in the economy consisting of NP^* pseudo-innovators and $N(1 - P^*)$ absorbers would exceed returns to N absorbers, and exceed, by even more, returns to $(N - 1)$ absorbers and one pseudo-innovator. Such a situation could only arise if there were external returns to scale in innovation sufficient to compensate for the extra internal cost to the firms.

How might external returns to scale in innovation come about? In the static context within which our game theoretic model exists we cannot rely upon learning for the presence of external returns to scale; the explanation will have to lie in structural or institutional factors. As examples, we can imagine three types of ancillary activities which facilitate innovation, one by making capital available, a second by providing specialised services and a third by assuring knowledgeable suppliers and customers. Capital is needed to finance research and development, which does not, initially at least, generate real assets. Yet in many developing countries money cannot be borrowed on the security of such an intangible asset as knowledge : it must be supported by physical or financial assets. But if capital is readily ventured on attempts at innovation the returns to the pseudo-innovators will be increased.

Similarly, the returns to pseudo-innovators will be increased if they can draw upon outsiders for services which they cannot supply economically within their own firms. There are (internal) returns to scale in such services as those provided by patent attorneys, testing laboratories, technical booksellers, market researchers, and process and equipment designers, returns which a single innovating firm is seldom able to exploit. A firm specializing in the provision of one of these services, and making it available to other firms carrying out research and development, can exploit the economies and consequently provide the service more cheaply.

The third ancillary factor generating external returns to scale is the existence of accommodating suppliers and customers for the innovating firm. The pseudo-innovator will be able to produce his novel product more cheaply if he can procure locally equipment designed and manufactured to his special needs. He will be able to sell it more profitably if local customers are accustomed to procure novel items from domestic firms.

For the above reasons it would not surprise us to discover that the total returns to all firms in the newly industrialized economy would be maximized with a goodly number of them acting as pseudo-innovators. But for numbers exceeding this optimal quantity, NP^*, total returns would be expected steadily to fall, as the costs to additional pseudo-innovators rise in line with the rising prices of the scarce factors used in research and development.

Given these features , the function $\pi(P)$ could take on the shape illustrated in Figure A, in which the returns to all pseudo-innovators are indicated on the horizontal axis and the returns to all absorbers on the vertical. Total returns in the economy are the algebraic sum of the two; its maximum is at $\pi(P^*)$, in the interior of the space. There is a local maximum at $\pi(P^0)$, and a local minimum at $\pi(P^-)$ between the two maxima. There is also another minimum, in all probability a global minimum, at $\pi(P^1)$ where all firms in the economy are pseudo-innovators.

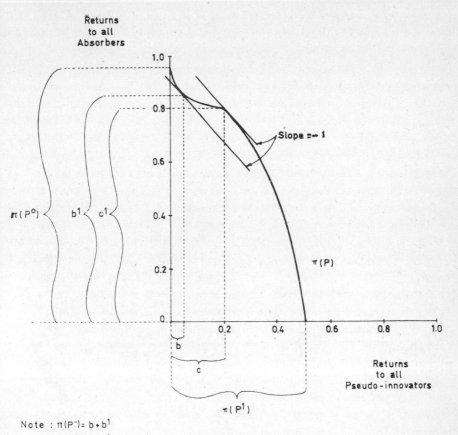

FIGURE A

The Overall Returns Function $\pi(P)$

At its interior maximum and minimum points in Figure A, the function $\pi(P)$ has a slope of minus one, this being the "trade off" for the individual firm between one unit of profit received from acting as a pseudo-innovator and one unit received from acting as an absorber. The firm's utility function assures that these are valued equally. The slopes of the returns function at the boundaries will both be less than minus one.

Although we have postulated the shape of the returns function from general arguments of an economic nature, in a game theoretic framework, the returns function will be derived from the pay-off kernel $K(x,y)$. We must ask what sort of pay-off kernel would generate the function plotted in Figure A? As drawn the function in Figure A could be represented by a polynomial of degree three; does this enable us to represent the pay-off functions as poly-nomials also? It would be helpful if it did, since infinite games on the unit interval whose kernels can be expressed as polynomials are fairly tractable, the properties of their solutions being generally known and methods for their numerical computation being available. Unfortunately, as can be seen from equation (2), the overall returns function is a linear combination, with weights P and $(1 - P)$, of the expected pay-offs, which are themselves linear combina-tions of the elements in the pay-off matrix. Given these links, and starting with a specified returns function $\pi(P)$, we can only deduce the shape of the kernel function $K(x,y)$ if the returns function is convex throughout, or if it is concave throughout. But the returns function in Figure A is convex over part of its range and concave over the remainder; as a result it is not possible to determine the properties of its inverse, which is the kernel function.

It should be possible, though, to conjure up a mutually consistent set of play-off matrices for the labelled values $\pi(P^0)$, $\pi(P^-)$, $\pi(P^*)$ and $\pi(P^1)$ (equi-valent respectively to P^0, P^-, P^* and P^1) in Figure A. These can then be com-pared in order to see if the changes in the four elements in the cells, the $a_{i,j}(P)$ are in the expected direction. The four matrices are presented in Table C.

We commence with the pay-off matrix which generates the highest re-turns for all the firms in the economy, labelled P^*. Total returns are unity, with the returns for any single firm at each encounter with a rival being 3, $1/2$, $5/9$ or $10/9$, depending upon the firm's choice of role and upon which type of adversary — pseudo-innovator or absorber — it meets. When the structure of the economy is that indicated by P^* — equivalent to twenty per cent of all firms being pseudo-innovators and eighty per cent absorbers — the expected returns to the single firm will be equal, whichever role — pseudo-innovator or absorber — it chooses. This equality can be calculated simply by fitting the appropriate numbers from the pay-off matrix for P^* into equa-tions (1a) and (1b).

A similar equality of expected returns from the alternative choices exists at P^-, where total returns to all firms are 0.9 units. At P^- the economy con-sists of five per cent pseudo-innovators and ninety-five per cent absorbers, and is in an unstable equilibrium, in the sense that any slight departure from the proportions five per cent/ninety-five per cent would lead, via a change in the numbers in the pay-off matrix, to a cumulative movement towards either P^* or P^0. Each firm would discover, when comparing the expected returns from alternative strategies, that one is higher than the other. It would therefore be preferable to choose to be of one, rather than the other type. It would not be until the overall structure shifted to either P^* or P^0 that stability would have been attained.

TABLE C
Pay-off Matrices Consistent with Overall Returns Function at Points P^0, P^-, P^* and P^1

At P^0
$(\pi = 0.95)$

	Y_1	Y_2
x_1	$3 < a_{1,1} < 10$	$< \dfrac{19}{20}$
x_2	$\dfrac{71}{100} < a_{2,1} < 3$	$\dfrac{19}{20}$

At P^-
$(\pi = 0.9)$

	Y_1	Y_2
x_1	10	$\dfrac{8}{19}$
x_2	$\dfrac{71}{100}$	$\dfrac{91}{100}$

At P^*
$(\pi = 1.0)$

	Y_1	Y_2
x_1	3	$\frac{1}{2}$
x_2	$\dfrac{5}{9}$	$\dfrac{10}{9}$

At P^1
$(\pi = 0.5)$

	Y_1	Y_2
x_1	$\frac{1}{2}$	$> \frac{1}{2}$
x_2	$> \frac{1}{2}$	$> \frac{1}{2}$

Although P^0 on the boundary where all the firms in the economy are absorbers is a stable point, it differs from P^* and P^- in that the expected returns from being an absorber exceed those from being a pseudo-innovator. The pay-off matrix for P^0 reveals this inequality; at P^0 all the rivals that the ith firm encounters are absorbers, so only the second column is relevant to the firm's choice. If the firm chooses to be an absorber too, it will expect to receive 19/20ths; if it chooses to be a pseudo-innovator, it will expect to receive less than 19/20ths. Its rational choice is to be an absorber, like all the rest.

An identical choice would be made by the rational firm at P^1, as indicated in the fourth of the pay-off matrices. In an economy consisting entirely of pseudo-innovators, the ith firm finds its expected returns to be higher if it chooses the other role. In this case, the nearest stable point where the expected returns from either choice are equal is P^*.

The four pay-off matrices appropriate for P^*, P^-, P^0 and P^1 are consistent with the aggregate returns function illustrated in Figure A. Functions $a_{i,j}(P)$ of which the four pay-off matrices are particular points and which are consistent with the entire function $\pi(P)$ can be found by trial and error. These four functions could be plotted over the range of P, $0 \leqslant P \leqslant 1$; such plots would be of interest to the games theorist and the mathematician, whose analyses commence with the kernel function. To the economist, however, possibly more interesting plots would be those of Figure B. Like the $a_{i,j}(P)$ functions, the expectations functions of Figure B are plotted over the domain of the structural variable P, but each $a_{i,j}(P)$ is weighted by either P or $(1 - P)$ according to equations (1a) and (1b). At the extreme points of the domain of P, one or the other term dominates, but in the middle of the domain the two terms in the expectation functions both count. The advantage to the economist of plotting the expectations functions is that they include in a single plot both the pay-off entries to the single firm (the $a_{i,j}(P)$) and the distribution of its rivals by type of firm (the P). The plots in Figure B thus reflect the firm's competitive environment, which is compounded of the types of encounters it has and the frequency with which different types arise.

Before we move on to drawing the implications from the game-theoretic analysis of competition among different types of firms in an economy we should mention that the analysis can be generalized to any number of types. So long as we continue to express the game as infinite, with continuous pay-off kernels, defined over the unit interval, we can be assured that solutions equivalent to P^*, or to P^0 or P^1, or to both, will exist. Difficulties there will be, of course, but they will not be because of the dimensionality of the problem; as so often in economics they will be because of the unavailability of data or because of the inappropriateness of our game as an abstraction of the competitive environment in a developing country.

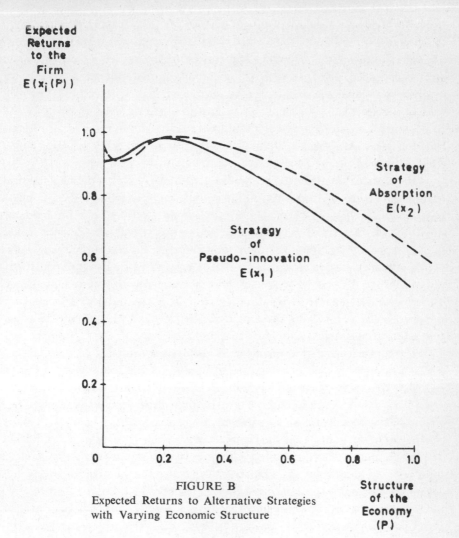

FIGURE B
Expected Returns to Alternative Strategies
with Varying Economic Structure

Implications of the Game-Theoretic Analysis

No one believes that theory is neutral. Games theory, like any other, has implicit within it presumptions about the nature of economic activity : about who the economic agents are, about what their objectives are, about what information they have and to what use they put it, and about how they interact with other economic agents. These presumptions shape the questions that are asked, the research that is carried out, and the implications for behaviour. In this final section of the essay we will chiefly be drawing the implications for public policy in developing countries of the game-theoretic analysis of economic structure; but beforehand we need to consider the accuracy of our description of the economic environment. The results of the previous section depended, amongst other things, upon our assumptions as to the way in which individual agents interact (in successive, pair-wise contests between randomly-selected firms) and as to the pay-offs received by each of the contestants (the $a_{i,j}(P)$, or their familiar transformation, the overall returns function $\pi(P)$). Are these realistic? If not, is there any other criterion which would justify their use?

The first assumption (of pair-wise contests) is akin to the neo-classical market structure of duopoly, as has long been recognized by economists (e.g. Shubik (1959), Osborne (1971), Friedman (1977)). That an industry might be characterized as a duopoly, therefore, needs no defence other than an appeal to custom. The question is whether or not an industry comprised of more than two firms can be characterized as a single set of transitory duopolies. For this to be true each transaction with a customer of the industry or a supplier to it would involve two of the industry's firms in a contest : contests involving three or more firms would be excluded, as would an exclusive contract to supply or to purchase.

The alternative characterization of an industry implied in neo-classical micro-economics is one of competition among all the firms comprising an industry for every purchase or sale. In a duopoly this would involve the same two firms in ever-repeated contests; in a duodecapoly the same twelve firms; in a centopoly the same hundred firms. Taking the industry with many participants, we doubt that all firms contest every purchase and sale; but within the context of a developing country we also doubt that their modern industries contain many participants. Typically, the number of firms lies within the range of one to a dozen. Say the number is a dozen : then the two alternatives, the neo-classical and the game-theoretic, are that each sale or purchase is contested by the entire twelve firms − the neo-classical description − or that each sale or purchase is contested by two out of the twelve firms, the particular two arising by chance − the game-theoretic description.

These two alternatives are over-simplifications, of course, and both neo-classicists and games-theorists have devised more realistic analogues to the organization of industries, chiefly by allowing for the formation of cartels or coalitions; but these analogues are so complex that their general equilibrium properties cannot be deduced. We have no idea, in principle, how an economy of cartels, or of industrial coalitions, would behave. Since our purpose in this essay is to say something about the general equilibrium properties of an economic system with different types of agents, we would be advised to retain the simplest characterization.

That still leaves us with the question of which of the two simplest alternatives is the more accurate, and of little in the way of argument or evidence to lead us to prefer one or the other. To be sure, the Walrasian general equilibrium model has splendid properties, all of which are known in principle and some of which could be calculated for specific situations, if those situations could be properly described. As we are attempting to display in this essay, our particular game-theoretic model also has certain general-equilibrium properties, which could also be calculated if the specific situation could be properly described. In the case of the Walrasian model, proper description consists, in part, of all the production functions and the utility functions; in the case of our game-theoretic model, of the pay-off kernel — i.e. of the functions $a_{i,j}(P)$, where i and j would extend over all the types of economic agents in the economy.

With knowledge of the $a_{i,j}(P)$ the economist employing the model of this essay can determine the overall returns function of Figure A. He would need ancillary data to be able to determine inputs, outputs and prices, the wherewithal of the Walrasian; but he would already have calculated such interesting points as P^* and P^-, to be compared with the current structure, P^{actual}. To the economist working in a developing country its structure is as important a matter as its output and prices.

To the theorist, the plots in Figures A and B have an additional interest, in that they are a formalization of the thesis of the "big push" in developing countries. Imagine that the $a_{i,j}(P)$ underlying the two figures are accurate and that the current structure of the economy is reflected in a value of P less than P^-; i.e. $0 \leqslant P < P^-$. In such an economy any firm contemplating undertaking research and development would find that the expected returns from that choice would be less than the expected returns from choosing to be an absorber. The equilibrium distribution of rational firms would consequently be at P^0, where all firms are absorbers. In order to reach equilibrium at P^*, where there are a positive number of pseudo-innovators, there would have to be some collective effort of major proportions to alter the structure of the economy so that the number of pseudo-innovators would exceed NP^-. Once the number NP^- had been exceeded, and given the pay-off functions $a_{i,j}(P)$ postulated, enough firms, acting independently in their own self-interest, would decide to be pseudo-innovators to bring the economy to P^*. The initial, collective effort altering the structure of the economy from P^0 to beyond P^- would constitute the "big push".

Just how big the push would have to be would depend in practice upon the difficulty involved in setting up the institutions generating external economies and, if taxes or subsidies to firms are to be used to achieve their compliance, to the total amounts of inducements. In principle, the difficulty and the inducements would be measured by the differences in the pay-offs to firms between P^0 and a point slightly beyond P^-, multiplied by the number of firms converted from absorbers to pseudo-innovators NP^-. The pay-off function $a_{i,j}(P)$, evaluated at the appropriate points, would provide the data needed to calculate the magnitude of the impetus.

The frequent reference to the pay-off functions gives some indication of their importance in a game-theoretic analysis of industrial structure. Together with some understanding of the way in which firms interact (in pair-wise contests or otherwise), the estimation of the $a_{i,j}(P)$ would be on the agenda of any research programme. To estimate the $a_{i,j}(P)$ over the entire domain of the variable P would clearly be impossible, but it might be possible to record the values within the neighborhood of P^{actual}, reflecting the current economic environment, and to estimate the values at P^- and P^*.

Such a programme presumes that the current structure of the economy is represented by P less than or equal to P^-, $(0 \leqslant P^{actual} \leqslant P^-)$, and that the desired structure is represented by P^*. The latter presumption leads us into the area of public policy, and the final section of this essay.

Public Policy within a Game-Theoretic Framework

Within the framework of games theory, public policy is directed towards individual firms in such a way as to influence their interactions with other firms and the rewards they receive. The chief objective of public policy, for which influence is exerted, is an ideal organization of industry.

In setting up our game-theoretic model we subsumed one mode of interaction, namely thet of pair-wise competition between different firms in every transaction. Moreover, we subsumed that the rewards to firms were generated solely through these transactions, and that they took the form of amounts of money equally valuable to all. We could have considered other forms of interaction, for example a corporate system in which a coalition of firms rather than each individual firm was the economic agent, but this would have made the analysis much more difficult; we could have considered other forms of the utility function, but any less stringent form would have made the analysis more difficult. We could have considered other objectives besides industrial structure, but this would have extended the model beyond the (mathematically tractable) unit interval.

If it seems that what was left to consider is too limited to be of any consequence we can cite some advantages of the game-theoretic approach. First of all, attention is directed towards individual firms at the point where their behaviour is most evident, where they interact with other firms. Attention is directed to the market.

Secondly, attention focuses on phenomena that are already under observation, namely monetary rewards. The numbers that appear in the cells of the pay-off matrix are average returns per transaction, i.e. unit profits, the concerns of tax authorities : no attention need necessarily be paid to matters internal to the firm such as conditions of cost.

Thirdly, given that the focus of attention is individual firms in markets the "towards whom" and "where" of government intervention have been determined. With this knowledge it is easier to determine the "how" of intervention. The data with which the game-theoretic model operates suggests that intervention take the form of altering the pay-offs of encounters with rivals, according to the type of firm. Governments of developing countries already intervene in this fashion, setting market prices or margins, allocating scarce resources, rewarding conspicuously good behaviour and penalizing bad : policy is already firm-specific.

The novelty in altering pay-offs arises from their units of measure. The dimensions of pay-off are so many units of currency, on the average, per encounter between two firms of specified type. There is no novelty in the numerator, which can be averaged over a string of encounters covering, say, a fiscal year. The novelty arises out of the dimension of the denominator, which involves the types of both contestants. Profits are customarily reported by individual firms as an aggregate over contests with *all* firms encountered, whatever their type. To alter pay-offs systematically, the government would have to collect from firms information on the outcomes of transactions according to the type of contestant. In the simple example that we have used in this essay there were only two types of contestants, pseudo-innovators and absorbers, and, consequently, only four possible outcomes. Had there been three types of firms (say pseudo-innovators, absorbers and those who employed a technique traditional to the developing country) there would have been nine possible outcomes; had there been n, n^2 possible outcomes. The data requirements are therefore identical to those of input-output analysis; one scalar in each cell if constant coefficients (the $a_{i,j}$) are assumed, or one function in each cell (the $a_{i,j}(P)$) if variations with regard to structure are to be allowed for. However, these are data that cannot be abstracted from ordinary accounting records, so the government which attempted to alter the pay-offs from individual contests would find it difficult to estimate pay-offs in advance of intervention.

Putting aside difficulties in implementation, let us see what policy a government guided by our game-theoretic model might adopt in the simple case with only two types of firms. Referring to the example in this essay, let us imagine that the developing country has yet to undertake any research and development of its own; i.e. $P^{actual} = 0$. Let us assume that the government is determined to foster innovation, with the aim of achieving an industrial structure containing a proportion, P^g, of all firms as pseudo-innovators.

The initial and final (= desired) points are only part of the information government needs in order to formulate its policy; also neccessary is information about the shape of the overall returns function and its underlying pay-off kernel function. Let us assume that the functions summarized in Figures A and B are apt, and that the final point, P^g, lies beyond P^*. This is equivalent to desiring to carry out more research and development, in lines both civilian and military, than that amount which would maximize returns for the firms in the economy.

For P^g to be stable, the expected returns to the firm facing a choice between pseudo-innovation and absorption would have to be equal; in the terminology of the preceding sections $E(x_1)$ must equal $E(x_2)$. Since this equality would not exist in freely operating markets (only at P^- and at P^* are expected returns equal), government would have to intervene, altering the pay-offs. In our simple case, there are four different pay-offs with which the government can tamper, $a_{1,1}$, $a_{1,2}$, $a_{2,1}$ and $a_{2,2}$. In order to equalize expected returns at P^g greater than P^*, the first two pay-offs should be increased, or the second two decreased, or both pairs should be changed, the first pair being increased relative to the second.

To the government, tampering with the first pair would probably be easier than tampering with the second pair because there would be fewer pseudo-innovators than absorbers within the economy and consequently fewer firms' returns to alter. The government could procede firm by firm, encouraging specific firms to become pseudo-innovators. The changes in returns to all firms will be the same, of course, regardless of whether it is the returns to individual pseudo-innovators or to absorbers that are altered.

Let us imagine that it is the expected returns to pseudo-innovation that are to be altered so as to achieve the structure indicated by P^g. It will not be sufficient for the government to equalize $E(x_1)$ to $E(x_2)$ at P^g because there may be other structures P at which the same equality would hold. Given the overall returns function illustrated in Figure A some structure between P^0 and P^- might also be sustained by the same set of pay-offs. If, as is likely, the economy is initially at or very near P^0, equalizing the expected returns to individual firms at the value appropriate to P^g would only bring the economy to this Γ, $(\Gamma^0 < \Gamma < \Gamma)$, with a smaller fraction of pseudo-innovators than desired.

It is to oversweep this "structural trap" that the "big push" is called for. Altering pay-offs is not a sufficient policy when there are multiple solutions supported by a single set of entries in the pay-off matrix. In our case, the big push is necessary to dislodge the economy out of the domain of P between P^0 and P^- into the domain beyond P^-. The push will have two components, the first being the creation of the institutions which generate external economies in research and development and the second being the persuasion of a sufficient number of firms ($NP^- + 1$ at least) to adopt the role of pseudo-innovator.

Once there are at least $NP^- + 1$ firms acting as pseudo-innovators, the self-interest of individual firms could be relied upon to bring the economy to P^*, but not to P^g where the latter exceeds the former. It is at P^* that the pay-offs need to be altered. One appropriate policy is therefore a sequential programme requiring initially the deliberate conversion of a sufficient number of firms into pseudo-innovators to oversweep the structural gap, then the voluntary shift of more firms until the total number of pseudo-innovators stabilizes at NP^*, and finally the alteration of pay-offs to bring the economy's structure to P^g . The government's role is initially active (converting $NP^- + 1$ firms), then passive (allowing the structure to seek P^*) and finally active again (altering pay-offs so that $E(x_1)/P^g)$ equals $E(x_2/P^g)$.

Another appropriate policy for the government would be to create the proper environment in one fell swoop. In our case, such a policy would require, simultaneously, converting NP^* (rather than the smaller number $NP^- + 1$) firms into pseudo-innovators and altering pay-offs so that the economy would settle at P^g. A third, and still more ambitious, policy would be to convert NP^g firms into pseudo-innovators and alter pay-offs so as to sustain this solution.

These three policies would achieve the same desired outcome but would, presumably, incur different costs. Some idea of the costs can be obtained by reference to the pay-off functions. Keeping the assumption that is the pay-offs associated with the first strategy (choosing to be a pseudo-innovator) that are to be altered, the appropriate data are available in Table C. The initial comparison involves the $NP^- + 1$ firms which are to be converted from absorbers, at P^0 the more attractive strategy, to pseudo-innovators. At P^0, $a_{2,2}(P^0)$ exceeds $a_{1,2}(P^0)$ by $19/20 - 16/20$ or $3/20$; subsidizing all $NP^- + 1$ firms by this amount would cost the government $(3/20)$ $(NP^- + 1)$ units of currency. Returns to all the firms in the economy would have fallen, as shown by the overall returns function in Figure A, from 0.95 units at

P^0 to just a little more than 0.90; the distribution of returns among pseudo-innovators and absorbers would be 0.0075 (with N normalized to unity and P^- equal to 0.05) and 0.8925 (0.90 minus 0.0075) respectively. The total cost to the government of converting five per cent of firms to pseudo-innovators would therefore be a little less than one per cent (0.0075) divided by 0.90 times 100 per cent) of the returns received by all firms; the total returns to all firms in the economy would fall by approximately four per cent (one per cent subsidy less the 5.0 per cent reduction in overall returns). Given our fictitious returns function, and underlying pay-off functions, the force behind the big push is provided by the unconverted firms, in the form of lower returns to each.

The second stage of the government's programme is one of inaction, while the structure of the economy adjusts to the stable point P^* through the voluntary response of individual firms to the changed pay-offs. At values of P exceeding P^- ($P^- < P < P^*$) the expected returns from pseudo-innovation are greater than those from absorption, so enough individual firms will choose this strategy to bring the economy to P^*. Since we have made the common assumption that there are no costs in adopting strategies, either to individual firms or to the economy as a whole, the second stage of the programme imposes no burden whatsoever.

In the third and final stage the government adopts an active role again, altering pay-offs so as to transform P^g into a stable solution. So as to derive quantitative results, let us take P^g as 0.25, exceeding P^* by 0.05 and yielding an economic structure with twenty-five per cent more pseudo-innovators than would exist were the government not to intervene. From Figure B we can see by how much the expected returns to pseudo-innovation would have to increase to make individual firms indifferent to the two alternative strategies. Reading off the distance between the two curves at $P^g = 0.25$, we find that an increase of 0.01 monetary units is needed per firm. Assuming that the government provides the appropriate subsidy, the cost will be 0.0025 units (equal to 0.01 per firm, multiplied by 0.25, the desired number of innovating firms). In the absence of government intervention the total returns to all firms at P^g would have been 0.9875; with government subsidies this figure is raised to 0.99 (equal to the intercept of $E(x_2)$ on the vertical axis of Figure B). The total amount of the subsidy is therefore 0.0025 units, and this must be provided constantly, or else the structure of the economy will revert to P^* ($= 0.20$).

It is now possible to state the total costs to the government and to all firms of achieving the desired economic structure by means of the three-stage programme. To the government, the cost of the first stage, the big push, is a once-and-for-all 0.0075 monetary units; of the second stage, nothing; and of the third stage, a recurrent cost of 0.0025 units. To the firms, the cost of the first stage, in terms of returns foregone, is approximately 0.04 units. These are more than recovered during the second stage when total returns increased from 0.90 to 1.00 units. The net change over the first and second stages is thus a positive 0.06 units ($-0.04 + 0.10$), and this occurs once and for all. In the third stage, all firms suffer a recurrent reduction in returns, relative to their position at the end of the second stage, of 0.01 units.

The sequential programme is only one of three alternative government policies, the other two being to convert $P*$ firms into pseudo-innovators, or to convert P^g, and thereafter alter pay-offs accordingly. Without going into the calculations we can state that the costs to the government are as follows : for the second policy, a once-and-for-all cost of 0.03 units and recurrent costs of 0.0025 units; and for the third policy a once-and-for-all cost of 0.0375 units and the same recurrent costs. The difference in the costs of the three alternatives arises because of the different number of firms being initially converted to pseudo-innovators through subsidies (higher in the third alternative than in the second, and in the second than in the first); the recurrent costs are the same for all alternatives. The number of firms converted to pseudo-innovators gives an indication of the magnitute of the initial impetus; the third alterantive requires the biggest push, the first the smallest.

The first alternative programme relies upon market forces, operating through the contests between different types of firms, to change the structure of the economy from P^- to $P*$, from $P*$ to P^g, and to maintain it at P^g. The second alternative relies upon market forces to change the structure from $P*$ to P^g and to maintain it there. The third alternative does not rely upon market forces to change the structure of the economy, only to maintain it at P^g. In our game-theoretic model, as in some others, reliance upon market forces, when they operate in a favourable direction, reduces the costs to government of achieving its desired outcome. What games theory does is to indicate over what ranges market forces can be expected to operate in the right direction and how much it would cost to over-ride market forces, through direct intervention. Games theory can also indicate where market forces, if left alone, can operate in the wrong direction; in our example this would occur over the domain of P between P^0 and P^-. Unless pay-offs were altered, an economy initially at a value of P slightly less than P^- would, through the choices of rational firms competing in free markets, stabilize at P^0, where innovation is completely absent.

In spite of their appearance, none of the results of the analysis above have any dynamic content; they are all of the nature of comparative statics.

The comparisons are between one equilibrium solution and another, saying nothing about the passage from one equilibrium to the next. In principle, it should be possible to formulate the problem stated in this essay in dynamic terms and solve it using optimal control theory. To do so, it would be necessary to determine the variability of the pay-off functions with time. In our simple case we would need four functions $a_{i,j}(P,t)$; in an n-dimensional case we would need n^2 time-dependent functions. Moreover, if the objective function of the optimal control problem contained any variables other than time itself (say, costs) we would need information relating changes in structure and in time to the costs of achieving these changes.

Let us conclude by summarizing the advantages and disadvantages of a game-theoretic approach to the analysis and direction of an economy with different types of firms. Most national plans lay out two sorts of objectives, structural and productive. In our game-theoretic approach, it is structure that is focused upon. The proper outputs are assumed to flow automatically from an economy with the proper structure. Structure determines supply, and supply is the bottleneck : not a wholly inaccurate characterization of the typically underdeveloped country. Not only does the game theoretic model abstract from demand but it also abstracts from the conditions of cost. Technology is considered only as it leads to the differentiation of firms into various types, and productive relationships are not considered at all.

What is focused upon is the competitive environment, where firms are attempting to maximize their returns while encountering like-minded rivals. The fundamental data with which the theory works are numbers representing the (average) return to each type of firm following each encounter. Returns are assumed to vary with the structure of the economy, structure being measured by the proportions of the various types of firms in the whole. It was argued, almost out of total ignorance, that the game-theoretic view of the generation of rewards in an economy — as an outcome of pair-wise contests between producing firms — was as realistic as the views underlying different bodies of economic theory.

Viewing economic activity in this way, it was possible to employ a particularly tractable branch of games theory — infinite games with continuous symmetric kernel functions played over the unit interval. The existence of solutions to such games has been proven. The economic environment conforms to these games, provided that firms have non Neumann-Morgenstern utility functions, are aware of the average returns to each contest, given their own and their rival's choice, and adopt a mini-max/maxi-min strategy. A simple example, involving only two types of firms — pseudo-innovators and absorbers — was concocted so as to illustrate possible solutions and the pay-off functions that generated them. The particular pay-off functions employed were consistent with an overall returns function for the universe of firms similar to the production possibility curve of international trade theory.

By employing individual pay-off functions which yielded an overall returns function exhibiting, in sequence, decreasing, increasing and again decreasing economies of scale in one of the two types of activities we were able to illustrate three equilibrium solutions (two stable and one unstable). The first could be thought of as characterizing the economy of a developing country which undertakes no research and development; the third as characterizing the same economy after various external economies, aiding research and development, have been achieved. The second, unstable equilibrium would lie between the two stable equilibria, characterizing an economy which had invested in some, but not enough, of the ancillary activities generating the external economies.

Comparisons of the pay-off matrices that underlay these equilibria provided examples of the effects and costs of alternative government policies. Costs were measured in the same units as returns, and divided into those borne by the government and those by firms. In one case investigated, three alternative government policies were compared; the cheapest was the one which made the greatest use of market forces, where they pointed in the right direction.

Theory depends upon its realism, and policy upon its statistical accuracy. To assure the last there is required a knowledge of the pay-off functions, the $a_{i,j}(P)$ of the model. These, it will be remembered, are the average return to a firm of the ith type after an encounter with a rival of the jth type (where i,j extend over all types) within an economy of structure P. To say the least, such knowledge does not exist currently for any economy; whether or not it could be attained, over the relevant domain of P, will not be known until someone has tried to gather it. Such a task would seem to be worth the attempt.

The immediate advantage of acquiring knowledge of the pay-off functions for firms of each type would be similar to acquiring knowledge of supply and demand curves — understanding the behaviour of firms and industries and explaining rewards. Knowing pay-off functions, one would have the subsequent advantage of being able to employ games theory in carrying out analysis or in making calculations. And the game-theoretic model in this essay permits both analysis and calculation, within a régime where overall objectives are being attained while individual economic agents are following their own self-interest.

REFERENCES

ENOS, J.L. and PARK, W.H. (forthcoming), *The Absorption and Diffusion of Imported Technology : the Case of Korea*, Croom-Helm.

FRIEDMAN, J.W. (1977), *Oligopoly and the Theory of Games*, Amsterdam, North Holland.

KARLIN, S. (1959), "The Theory of Infinite Games", vol. 2, *Mathematical Methods and Theory in Games, Programming and Economics*, Reading, Massachusetts, Addison-Wesley.

MANSFIELD, E., SCHWARTZ, M. and WAGNER, S. (1981), "Imitation Costs and Patents: An Empirical Study", *Economic Journal*, vol. 91, December, pp. 907-918.

MAYNARD SMITH, J. (1982), *Evolution and the Theory of Games*, Cambridge University Press.

OSBORNE, D.K. (1971), "The Duopoly Game : Output Variations", *American Economic Review*, vol. 61(4), September, pp. 538-60.

ROCHFORD, S.C. (forthcoming), "Symmetrically Pairwise-Bargained Allocations in an Assignment Market", *Journal of Economic Theory*.

SHUBIK, M. (1959), *Strategy and Market Structure : Competition, Oligopoly and the Theory of Games*, New York, Wiley.

——, (1982), *Game Theory in the Social Sciences*, Cambridge, Massachusetts, The M.I.T. Press.

CHAPTER EIGHT

PUBLIC POLICY TOWARD TECHNICAL CHANGE IN AGRICULTURE

By Peter B.R. Hazell* and Jock R. Anderson**

INTRODUCTION

Since 1950, world agricultural output has grown at a more rapid rate than at any earlier time in history. Cereal production, for example, which accounts for over 90 per cent of the world's major food staples has grown at an average annual rate of nearly 3 per cent. Although individual country experiences have varied considerably, the developing countries as a group have performed as well as the industrialised countries (Paulino 1984).

Many factors have contributed to this spurt in agricultural output. These include an expansion of the cropped area into new lands; substantial investments in irrigation, rural electrification, roads and other rural infrastructure; investments in rural education and farm extension; price support policies and huge public subsidies to agriculture in many industrialised countries; the increased use of fertilisers and pesticides; and technological advances in the form of improved plant varieties, livestock, machines and management.

Despite the apparent success of world agriculture in recent decades, there are causes for growing concern. First, world population has grown and continues to grow at a little over two per cent per annum. This leaves little margin for increases in food supplies per caput, and hence little scope for the widespread alleviation of poverty because of the increased food demands that income increases for the poor generate.

Second, agricultural growth rates are slowing down. Resources for the Future (1984) estimates that world agricultural output rose at 3.1 per cent per year in the 1950s, 2.6 per cent in the 1960s, and 2.2 per cent in the 1970s. Paulino (1984) estimates that the growth in world cereal production dropped from 3.5 per cent in the 1960s to 2.7 per cent between 1970 and 1977. The decline reflects reduced opportunities for expansion of the cropped area into new lands or for increases in the irrigated area. Future growth will have to

* International Food Policy Research Institute, Washington, D.C., USA.
** University of New England, Armidale, Australia.

depend increasingly on agricultural research and the development of yield-increasing technologies.

Third, where rapid agricultural growth has occured, it has tended to worsen income inequalities, both between regions and between households within regions. Contrary to early criticisms of technological change in cereal production since the mid-1960s (the so called Green Revolution), more recent evidence does not show that technological change necessarily makes the poor poorer. However, because technological changes tend to favour one resource over another, and because resources are never equally distributed amongst households, the gain from technological change has rarely been "fairly" distributed between the landed and the landless.

In this essay we review briefly the nature, importance, and historical record of technological change in agriculture, largely with a view to identifying factors that are important for continued technological change to meet future food needs. We also examine arguments and evidence about the distri-butional consequences of technological change, and consider microeconomic factors determining the adoption of new technologies. In conclusion, we consider the kinds of policies that are necessary for agricultural technology if future growth and equity goals are to be met.

AGRICULTURAL TECHNOLOGY AND ECONOMIC GROWTH

The Nature of Technological Change in Agriculture

Technological change results when farmers adopt new technologies. Its impact can be measured at an aggregate level through the estimation of indices or production functions which capture the change in output obtained with given bundles of inputs (Ruttan 1956). A simple and commonly used measure of technological progress is the average ratio of output per unit of of a scarce resource, for example, yield per hectare or yield per day of labour. Such partial measures are only strictly valid if the quantity and quality of all the other inputs used are held constant. There is a substantial literature on methods of measuring agricultural productivity under more realistic assumptions, (see Capalbo and Trang (1984) for a recent review) but this literature need not distract us here.

It is useful to define a technology as representing a particular parameterisation of a production function. Technological change is then defined as a change in the parameters of, or a structural shift in, the production function (Ruttan 1960). By this definition, a simple increase in the application of fertilizer is not a new technology whereas the use of an improved plant variety which increases the response to fertilizer is a new technology.

When defined in this way, virtually all technological change arises from changes in the quality of one or more inputs (such as improved genetic capability or land drainage) or from the introduction of entirely new inputs (such as some machines, or irrigation water in dryland areas). The cost of technological change is therefore embodied in the cost of inputs; there are no "manna from heaven" shifts in the production function other than transitory changes induced by weather.

Technological change in agriculture is most often embodied in capital inputs, for example land drainage and enclosure, irrigation, machines and buildings. Even the new varieties associated with the Green Revolution have typically required substantial and complementary investments in irrigation, drainage, roads and markets, in addition to the investment in agricultural research and extension which led to their creation and dissemination in the first place. When all these capital costs are added together, the incremental capital-output ratio for agricultural growth can be surprisingly high.

A distinguishing feature of agricultural technology is that a large part of the capital cost is borne by the government rather than the farmer. Most agricultural research and extension is universally provided as a public good, and investments in canal irrigation and rural infrastructure are typically funded by governments at little or no cost to the farmer. For this reason there can be important discrepancies between the cost of new technologies to farmers and the cost to society. There are also important distributional consequences in that many technologies are rendered scale neutral to the farmer whereas only the richer and larger scale farmers would benefit if the full costs had to be borne by the farmer.

That governments of many hues and colours are willing to subsidise technological change in agriculture in these ways reflects, in part, a universal concern with ensuring adequate food supplies for the urban populace. It also reflects difficulties inherent in privatising agricultural research and development (R & D) activities (a) which are too costly for individual farmers to bear, (b) which do not easily render a profit to the inventor because of difficulties in patenting or otherwise protecting the new technology, (c) which are costly to advertise and disseminate, particularly in developing countries, and (d) which can take many years to develop. There are, of course, some important exceptions. Private companies develop and sell a wide range of farm machinery and agrochemicals. Some plant and livestock breeding is also successfully carried out in the private sector, and this has been facilitated by the development of hybrids which cannot be reproduced by farmers. However, these activities are largely confined to the industrialised countries and, although the products are widely available through multinational corporations, they are not always appropriate to the needs and conditions of farmers in developing countries.

Most technological change in agriculture is biased towards saving labour (e.g., machines) or land (e.g., improved varieties). For economic efficiency, it is desired that the factor-saving bias of new technologies reflect the relative scarcity of factors in agriculture. Hayami and Ruttan (1971) provide evidence that this has, in fact, happened and that research in both the private and (more surprisingly) the public sector does respond to long-term trends in relative factor prices.

Agricultural Growth Linkages

Sustained economic growth requires an industrial revolution, but can a successful industrial revolution arise without there first being an agricultural revolution? This is a matter of some controversy amongst development economists, but what is more certain is that technological change in agriculture can be a great asset for national economic development, as well as an important source of employment.

Mellor and Lele (1973) and Lele and Mellor (1981) have argued the importance of increased food production in relaxing the wage-food constraint on economic growth. Mellor (1973) has also stressed the importance of technological change if agriculture is to provide the capital surpluses necessary to finance industrialisation. Further, by increasing farm incomes, technological change increases the purchasing power of rural households, thereby providing a large and vibrant market for the emerging industrial sector. This demand is reinforced by the input needs for a dynamic agriculture (e.g., fertilisers and machinery) and by the need for additional marketing, processing and transport services as production increases (Mellor 1976, Johnston and Kilby 1975, Quizon and Binswanger 1983).

The indirect impacts of agricultural growth can be substantial. In a study of agricultural and industrial performance in India, Rangarajan (1982) found that an 1 per cent addition to the agricultural growth rate stimulated about a 0.5 per cent addition to the growth rate of industrial output, and a 0.7 per cent addition to the growth rate of national income. At a regional level, Gibb (1974) found that each 1 per cent increase in agricultural income in the Nueva Ecija Province of Central Luzon in the Philippines generated an 1 to 2 per cent increase in employment in most sectors of the local nonfarm economy. Similarly, in a study of technological change in rice in the Muda region of Malaysia, Bell, Hazell, and Slade (1982) found that, for each dollar of income created directly in agriculture, an additional 80 cents of value added was created indirectly in the local nonfarm economy.

An important aspect of growth linkages to the nonfarm economy is that they are predominantly due to increases in household consumption expenditure. Bell, Hazell, and Slade (1982) report that about two-thirds of the 80 cent income multiplier in Muda was due to increased rural household demands for consumer goods and services; only one-third was due to agriculture's increased demands for inputs and processing, transport, and marketing services. Gibb (1974) also found strong employment links to the non-food consumer-oriented sectors in his study of Nueva Ecija. These findings support Mellor's (1976) contention that, because much of the accepted wisdom on development strategy ignores these consumption linkages, it has tended seriously to underestimate the potential importance of agriculture. Hirschman (1959), for example, in his influential study of the importance of linkages in promoting development, focused only on production linkages, and he found these to be weak for agriculture compared to most other sectors of the economy. On this basis, he recommended the greatest priority be given to public investment in nonagriculture.

In addition to enhancing agriculture's contribution to national economic growth, the existence of strong consumer expenditure linkages between agricultural households and the nonfarm economy is important for two other reasons. First, the income and employment generated by these linkages is predominately concentrated in rural areas. Rurally focused growth is desirable in many countries where rural areas have been severely disadvantaged in the past through urban-biased policies (Lipton 1977). Such policies have encouraged excessive migration from rural to urban areas and have exacerbated problems of rural underemployment.

Second, the kinds of goods and services demanded are typically produced by small labour-intensive enterprises. They are focused on such sectors as transportation, hotels and restaurants, entertainment, personal services, health, distributive trades, and housing and residential construction. Increased household demands for speciality agricultural products, particularly fresh fruits and vegetables, and fish and livestock products can also provide important increases in rural employment.

Strong household links to the rural nonfarm economy not only help alleviate problems of rural underemployment, but, because the major beneficiaries of the increased employment earnings are typically the poor, they also contribute to the reduction of rural poverty and malnutrition. Survey evidence from many countries confirms that the small farmers and landless workers obtain substantial shares of their total income from nonagricultural sources. Consequently, the beneficiaries of the indirect employment gains generated by agricultural growth need not be limited to poor, non-agricultural households residing in towns. Rather, they have the potential to touch a wide range of occupation groups within the poorer segments of society.

Historical Overview

From the beginning of settled agriculture some six thousand years ago until the middle of the eighteenth century, there was little change in the yields of most crops. Agricultural output kept pace with population growth largely as a result of increases in the cultivated area. New crops such as potatoes, maize and sugar beets also became important in parts of Europe.

Wheat yields increased slowly. Between 1300 and 1700, average wheat yields in England rose by only 50 per cent (Grigg 1984). They were still only about 0.5 t/ha in 1700. Wheat yields began to increase a little thereafter, facilitated by such developments as Jethro Tull's seed drill which, by planting the crop in straight rows permitted inter-row tillage; by the Rotherham mould board plough which permitted complete inversion of the soil and better weed control; by the incorporation of legumes and livestock into improved rotations; and by the development of artificial fertilisers in the 1840s (Deane 1965). By the 1850s, average wheat yields had risen to 1 t/ha in England, and had risen to 2 t/ha by 1909-1913 (Grigg 1984). They remained relatively static until the late 1930s, after which they increased exponentially. By the 1980s, average wheat yields in England had reached nearly 6 t/ha. Thus, in the space of 40 years, wheat yields increased by over 3 t/ha, or by more than twice the increase achieved in the whole of the preceeding six and one half centuries.

The story of English wheat is typical of the development of other food crops throughout the world. Progress everywhere prior to the Second World War was slow and patchy. Agricultural research and invention was largely eft to the individual farmer, and it lacked a rigorous scientific foundation. Colonial governments did make concerted efforts to increase the yields of important export crops such as cotton, rubber and tea, but no comparable public effort was made for food crops. The USA became a major exception when it established its Land Grant Colleges through the Hatch Act of 1887, but the full research and extension impact of these institutions did not occur until after the turn of the century (Hayami and Ruttan 1971).

Since the Second World War, world agriculture has undergone a major revolution on all fronts. Increased public expenditure on agricultural research, breakthroughs in genetics and in animal and plant breeding, and in the development of chemical fertilisers and pesticides, have permitted unprecedented growth in yields. They have also permitted farmers to specialise in the most profitable crops. No longer are farmers bound to fixed rotations and fallows to control weeds and pests and to maintain soil fertility. Rather, they are now free to practise increased monoculture and to plant the same crop year after year. In the industrialised countries there has also been a dramatic revolution in agricultural mechanisation. The horse, for example, has been entirely replaced by the tractor, and the combine harvester has eliminated the reaper and the thresher. This mechanical revolution was induced by rapidly rising wages as an ever increasing share of the workforce was absorbed by other sectors. In the USA, for example, the number of farms declined from 6.1 million in 1940 to only 2.4 million in 1980, while at the same time the value of agricultural output increased by about 80 per cent.

Third World Update

The most significant technological advance for developing countries occured with the development of high yielding cereal varieties (HYVs) in the 1960s — the Green Revolution. The first breakthrough occured with wheat in Mexico in the early 1960s with successful transfer to South Asia in the late 1960s, as a direct result of research supported by the Rockefeller Foundation. This was soon followed by the development of HYV rice at the International Rice Research Institute in the Philippines in the mid-1960s (Barker and Herdt 1982). The HYVs are not only favourably responsive to nitrogen but, because of their short straw, are not so susceptible to lodging when fertilised as are their traditional counterparts. When grown with fertiliser, pesticides, and adequate water, the HYVs can often produce more than double the yields of traditional varieties.

The impact of the Green Revolution on the agricultural production of developing countries has been substantial. The impact on wheat and rice production is a function of the area sown to the new wheat and rice varieties, and the increase in yields per unit of land. Increasing yields have made rice and wheat more profitable for farmers than certain other crops. Thus, in addition to yield increases on traditional wheat and rice land, more land has been brought into cultivation of these two crops. The HYVs have also facilitated significant expansion of irrigation and multiple cropping in many countries, thereby adding to the total area of these crops. Shorter growing periods and reduced photoperiodicity are important properties of the new varieties that have enabled increased multiple cropping.

It is estimated that between one-third and one-half of the rice areas in the developing world is grown with HYVs. In Latin America, for instance, about 2 million ha of rice is irrigated and another million grown under favourable rainfall and soil conditions but without irrigation (CIAT 1981). Estimated yield increases due to these varieties are some 1 t/ha in irrigated areas and 0.75 t/ha on favoured upland rice areas. Thus, total annual production increases in Latin America are around 2.5 Mt. Assuming a price of US 200/t of paddy rice, the value of the production increase for one year is about $ 500 million. Although these estimates are rough, it is clear that the production impact is large. Such a conclusion is supported by evidence from Asia. Herdt and Capule (1983) estimate that modern rice varieties annually add 27 Mt to the production of rice in eight Asian countries which produce 85 per cent of Asia's rice.

Earlier estimates of the production impact of modern rice varieties naturally were considerably below those reported above. The impact was estimated to be about 10 Mt in the Far East and less than 0.5 Mt in Latin America for the year 1976/77 (Pinstrup-Andersen 1982). While differences in estimation procedures explain some of the difference between the two sets of estimates, rapid increases in adoption since 1976/77 account for a large part, particularly in Latin America. Rapid increase in rice production due to modern varieties and associated inputs is still going on. In fact, except for Columbia, the bulk of the increase in Latin America has occured since the mid-1970s.

The wheat areas grown with modern varieties are of magnitudes similar to those for rice. It was estimated that about 30 million ha were grown with these varieties during 1976/77 (Pinstrup-Andersen 1982) and about 35 million today (James 1983). Modern varieties occupy a larger proportion of the wheat than the rice area. James (1983) estimates that the contribution of the new varieties to increased wheat production in developing countries was 7 Mt in 1982/83 worth about $ 1.2 billion. Earlier estimates for 1976/77 were 19.9 - 26.7 Mt (CIMMYT 1978) and around 21 Mt worth $ 2.5 billion (Pinstrup-Andersen 1982). The large differences among the estimates are due primarily to the different assumptions about average yield increases that are not known with a high degree of precision.

While the term Green Revolution originally described developments involving wheat and rice, high yielding varieties have been developed for a number of other food crops important to the developing countries. These include sorghum, maize, cassava, and beans. The area grown with improved maize varieties and hybrids derived from CIMMYT germplasm in developing countries reportedly runs into millions of hectares (James 1983). Major efforts to develop high yielding technologies for many other food crops grown under developing country conditions are more recent and attempts to estimate the global impact on production would be premature. However, evidence for some crops in a few countries, e.g., beans and cassava in Cuba and beans in various Central American countries, is encouraging.

Political and Economic Perspectives

Observers hold many and varied views of the nature and importance of technical change in agriculture (our review here parallels that of Hardaker, Anderson and Dillon 1984). Perhaps, since the mid-1960s, the predominant view is based on the notion that resource-poor farmers the world over behave rationally. Poor farmers in the developing countries are seen as being caught in a low productivity trap wherein, given the prices and factor availabilities they face, the methods available to them, and their limited capacities to take risks, there is little scope for them to reallocate their resources appreciably to increase their incomes (see, e.g., Stevens 1977). Given the seeming impracticality in many developing countries of materially moving prices in farmers' favour, the logical implication of the theory is that the low productivity trap can best be escaped by making agricultural methods available that are more productive than the existing set. The immediate policy prescription, therefore, is to strengthen research activities.

While this rationalisation of the importance of agricultural research and development is widely accepted (especially by involved researchers themselves), there are other views. One extreme example, Clayton (1983) argues that there is no justification for a specific orientation to meeting basic needs through income redistribution strategies such as emphasis in research on the development of "appropriate" technology for small scale farmers. Emphasis, he argued, should be on growth in productivity and output however it can be achieved. From this, small scale farmers will gain through the exercise of their natural enterprise, ability and energy. Another view is that, as developing countries move to having a majority of their population non-rural, food production for the urban majority will become a dominant concern. In consequence, research emphasis might then best be on high-input technology for medium to large-scale production with the welfare problems of small scale farmers being left to other policy instruments. Unfortunately, in most such developing countries (such as Brazil), welfare policies as such are virtually non-existent.

In the contrasting Marxist view, the particular techniques used in production are related to and are to some extent determined by the social organisation of production manifested largely in class relationships. There is, however, much disagreement among Marxists about the consequences of technical change. A conventional stance is to view technical advances within peasant agriculture as being likely to lead to differentiation of a class of capitalist farmers on the one hand, and a rural proletariat on the other (Lenin 1899). Thus it is seen as desirable to forestall these social changes by replacing backward peasantry by large state or communal farms that can be expected to reap advantages of scale and that should make it easier to extract the capital surplus required for industrial development. Chayanov (1925), however, argued on the basis of observed changes in Russian peasant agriculture that middle peasants were more resilient than Lenin had supposed, and that this resilience could hold up the hypothesised process of rural class differentiation. His theory of demographic differentiation saw the modernisation of peasant farming as being based on raising the technical level of agricultural production through agricultural extension and the development of cooperatives, within an institutional framework of family small holdings.

Elements of Chayanov's views are now to be found in the rural development strategies of several developing countries (e.g., Bangladesh and Indonesia) which have sought to preserve the structure of peasant farms but have encouraged the transformation of traditional agriculture through the introduction of improved methods by means of state-controlled agricultural cooperatives.

Many neo-Marxists see capitalism as the cause of poverty in the Third World, not as part of the solution (see e.g., Cole, Cameron and Edwards 1983). They view the causes of poverty in the periphery (notably the agricultural sectors of developing countries) as arising from the process of capital accumulation at the centre or core of the developed countries' economies (Sweezy 1968, Gurley 1975, de Janvry, 1977). The process of "unequal exchange" whereby the rich exploit the poor, is argued to oblige that most of the benefits of technical change in agriculture at the periphery will be captured by the central extractors of surplus value. The prescription implied is that political reform of the economic system, perhaps falling short of its revolutionary overthrow, must antecede use of technical innovations to attempt to reduce rural poverty. Analogous implications are held to obtain for the international generalisation of the unequal exchange school of thought (Prebisch 1959, Emmanuel 1969).

A further source of concern about the utility of improved production methods in reducing rural poverty arises from the recognition that, at least for agriculture producing a surplus sold in markets with inelastic demand, the main beneficiaries of innovations causing outward shifts in supply are consumers rather than producers. This neoclassical view thus can lead to conclusions somewhat similar to those of the neo-Marxists about the potential, at least under some market conditions, for using improved technology as an instrument to reduce rural poverty.

Market prices are central to those who argue that many problems of developing agriculture can be traced to distortions from free market prices. It is argued that technical and even institutional changes, will be induced in response to prevailing price and resource situations (Binswanger and Ruttan 1978). Although advocates of this theory recognise a need for some intervention in the processes of R & D, such intervention should be restricted to cases of market failure.

More extreme critical observers, such as Griffin (1979) and Lappé and Collins (1980), of aid in agricultural technology generation impute less than altruistic, if not overtly sinister, motives to some donors to the international research system with suggestions of powerful countries' desires for encouraging technological dependency, capitalistic imperialism and transnational profitability. We do not wish to understate the difficulties in endeavouring to target technology to benefit the economically and politically weakest members of agrarian societies (see Barker and Herdt 1982). However, until the Day of Freedom heralds the Perfect Society, we judge to be reasonable the present practice of technology enhancement agencies working with whatever (responsible?) authorities are in power, in order to do whatever can be done to reach and alleviate the lot of the weak.

DISTRIBUTIONAL CONSEQUENCES OF CHANGING AGRICULTURAL TECHNOLOGIES

It seems endemic to the nature of agrarian economies that they are destined to generate unfavourable distributions of households income. The classical economists, especiallly Ricardo, showed that, with a fixed endowment of land and limited technological change, continued population growth leads to the impoverishment of the labouring classes. As the pressure on land increases, cultivation expands into increasingly marginal areas, thereby increasing land rents and food prices, and reducing the average productivity of labour. If technological change is sufficiently weak in relation to population growth, the Mathusian prediction may be realised where labour's earnings in food equivalents fall so low that the population size eventually stabilises through increased mortality rates.

The classical theory was developed in England around 1800 at a time when the growth rates in both population and agricultural productivity were considerably less than the rates observed in today's developing countries. Nevertheless, the theory has potential relevance whenever growth in food production is inadequate. In an economy closed to international trade, the real price of food can only remain stable or decline if the growth rate in food production is at least as large as the growth rate in population plus the growth in per caput income times the income elasticity of demand for food. For most developing countries, this condition has required an agricultural growth rate in excess of 4 per cent per year since 1960. Few countries have achieved such growth rates on a sustained basis. In fact, during the period 1961 to 1977, the average annual growth rate in cereal production for the developing world was only about 2.9 per cent (Paulino 1984). Although respectable by any historical comparison, it is partly because the growth in food supply fell below the growth in demand in many countries that the dual phenomena of increasing agricultural productivity and worsening impoverishment of the labouring classes have often been observed.

Increasing rural impoverishment might be curtailed through rapid industrialisation and the international trading of manufactures for food. Such a strategy has rarely succeeded. The alternative hope is for substantial and sustained increases in agricultural productivity through technological change.

While technological change has the potential to reverse the classical trap of declining labour productivity, it may also set in motion new forces which act to worsen rural inequalities even as absolute poverty diminishes.

Individual improvements in agricultural technology tend to be specific to particular crops or environments. The breakthrough in HYV rice in the early 1960s, for example, was largely restricted to tropical areas having good rainfall or irrigation. Even today, HYV rice that is suitable for less favoured, dry-land zones has not been successfully developed. This specificity by crop and location of agricultural technology leads to disparities in the gains between countries, between regions within countries, and between farms. The diversity of annual growth rates in foodgrain production achieved by different states in India since the 1960s illustrates this point. These range from about 8.0 per cent in the Punjab to 1.99 per cent in Bihar for the period 1960/1 to 1978/9 (Sarma 1982). A particularly unfortunate feature is that the countries and regions that have benefited most from technological change in agriculture are often the most ecologically favoured, and hence the initially more prosperous. Plant breeders have yet to find ways of significantly increasing crop yields in the absence of generous amounts of each of the fundamental factors of plant nutrients, water and sunshine. The goal may well be unattainable.

Regional disparities within countries can be reduced through labour migration; a now common occurrence within many developing countries. But disparities between countries are more difficult to deal with, particularly where political boundaries prevent workers from moving between ecologically different zones. This is a particularly difficult problem for people in Africa.

The consequences of changing agricultural technology for the distribution of household income within rural areas are more complex, and have been the subject of a large and growing literature. This literature has predominantly focused on income distributional changes associated with the Green Revolution in the developing countries since the mid-1960s.

One of the attractions of the Green Revolution technologies is that they are, in principle, scale neutral, and can raise yields for small scale as well as for large scale farmers. Yet a number of early studies of the impact of the Green Revolution concluded that the rural poor did not receive a fair share of the benefits generated. It was argued that large scale farmers were the main adopters of the new technology, and smaller scale farmers were either unaffected or made worse off because the Green Revolution resulted in lower product prices, higher input prices, efforts by large farmers to increase rents or force tenants off the land, and attempts by larger scale farmers to increase land holdings by purchasing smaller farms, thus forcing those farmers into landlessness. It was also argued that the Green Revolution encouraged unnecessary mechanisation with a resulting reduction in rural employment. The net result, as argued by some, was a rapid increase in the inequality of income and asset distribution, and a worsening of rural poverty in areas affected by the Green Revolution (e.g., Griffin 1972, 1979, Fraenkel 1976, Harriss 1977, Hewitt 1976, ILO 1977).

The validity of these conclusions has not proved robust when subject to the scrutiny of more recent evidence. Ahluwalia (1985), for example, provides evidence that the incidence of rural poverty in India declined almost steadily between 1967/68 and 1977/78. This is contrary to the findings of Griffin and Ghose (1979) who analysed comparable data for the period 1960/61 to 1973/74. Ahluwalia (1978, 1985 forthcoming) also found that the incidence of rural poverty is negatively related to agricultural income levels per head. Bose (1982) showed that the proportion of the rural population in Indonesia falling below an asumed poverty level of consumption declined significantly between 1970 and 1976. By contrast national statistics on employment and wages suggest an increase in the incidence of poverty over this same period.

At the village level, Kikuchi and Hayami (1982) found that, in one Laguna village in the Philippines which benefited substantially from HYV rice, real wages remained approximately constant between 1960 and 1976 despite rapid population growth and in-migration. They compared these findings with a similar village in Java in which HYV rice failed to take. Real wages there declined between 1966 and 1976, in accordance with the Ricardian model, despite lower population growth.

Bell, Hazell and Slade (1982) also provide micro evidence that agricultural technology can help alleviate absolute rural poverty. They studied the combined impact of an irrigation project and HYV rice in the Muda River region of Malaysia over the period 1967 to 1974. The average per caput income of the population living in the project area increased by 70 per cent when measured in constant prices. Land owning households gained relatively most, but landless paddy workers also increased their real per caput incomes by 97 per cent, and this despite a wholesale shift to tractor mechanisation for land preparation. (Bell, Hazell and Slade, Table 7-7).

Where did the earlier studies err? Pinstrup-Andersen and Hazell (1984) offer four possible reasons. First, the studies were conducted too soon after the release of the Green Revolution technologies. While it was true that early adopters were primarily larger scale farmers, the studies failed to recognise that smaller scale farmers would follow quickly once they observed the success of their larger scale brethren. See, for example, the studies by Byerlee and Harrington (1983), Chaudhry (1982), Pinstrup-Andersen (1982), Blyn (1983), Herdt and Capule (1983) and Prahladachar (1983).

Second, the benefits to the poor, as consumers of rice and wheat through lower prices, were largely overlooked. Empirical evidence of consumer gains from technological change in developing country agriculture is plentiful (e.g., Akino and Hayami 1975, Mellor 1975, Evenson and Flores 1978, Scobie and Posada 1978, Pinstrup-Andersen 1979 and Scobie 1979). The consumer gains come about because food prices are lower than they would have been in the absence of the production increases induced by technological change. Population growth, import substitution, export, and domestic price policies can dampen the price reduction. In fact, price and foreign trade policies have been used extensively to strike a more desirable balance between the harmful effects of price falls on farmers and future food production, and the beneficial effects on consumers. Since the Green Revolution generates an economic surplus by more efficient utilization of resources and reduced unit costs, consumer gains need not imply producer losses. Both may gain.

Third, little or no attention was given to the multiplier effects of the Green Revolution, and the resulting impact on the incomes of the rural poor. Available evidence suggests that these indirect effects are about as important as the direct effects of technological change (Gibb 1974, Bell, Hazell and Slade 1982). The indirect benefits, however, are not restricted to the poor. They also increase the earnings of skilled workers as well as providing lucrative returns to capital and to managerial skills. In the Muda study, for example, Bell, Hazell and Slade found that the indirect benefits of the project were skewed in favour of the nonfarm households in the region, many of whom were relatively well off. They also found that, even among agricultural households, the landed households fared better than the landless. The point to be made is that, although the indirect effects of agricultural growth are unlikely to improve the relative distribution of income within rural areas, they can still have wide-reaching effects in alleviating absolute poverty.

Fourth, the impact of the Green Revolution was frequently confused with the impact of population growth, or with institutional arrangements, agricultural policies, and labour-saving mechanisation. Such confusion leads to incorrect identification of the causes of rural poverty, and thus to inappropriate recommendations for action to reduce such poverty. It also leads to a failure to appreciate the extent to which poverty and malnutrition would prevail today without the additional food bestowed by the Green Revolution.

The distributional consequences of agricultural technology are not immutable but can be influenced by public policy. Policies aimed at facilitating access to inputs, such as fertilisers and credit, can greatly reduce institutional and market biases against small scale farmers. Public investment in rural infrastructure, markets and integrated rural development can also promote technological change and output expansion among small-scale farmers, particularly in more remote areas (Pinstrup-Andersen 1982). The design of technologies themselves, can also be directed to benefit smaller scale farmers. For example, research can be focused on scale-neutral technologies, or on the crop and farming systems of small scale farmers (Anderson and Hardaker 1979). Last, but by no means least, land reform measures can reduce the skewed distribution of ownership of land, the factor which is probably most responsible for the bias in the distribution of the gains from technological change. Unfortunately, few governments have had the political will or clout to implement significant changes in the distruibtion of land. But it is these kinds of failings that lead to increasing rural inequalities rather than the new technologies themselves.

MICROECONOMICS OF TECHNICAL CHANGE IN AGRICULTURE

Agriculture is never static and, as noted in our historical commentary, has in some parts of the world, recently been positively dynamic. Farmers everywhere continually experiment, albeit informally, with new ways of doing things in the hope of finding improvements or as Warren and Livermore

(1911, p. 385) aptly put it long ago, "Every farm is an experiment station and every farmer the director thereof". With the limited resource situation of most farmers, especially in the Third World, and the well known difficulties of capturing benefits of most agricultural innovations, public investment in agricultural research has a long tradition of importance relative to work in the private sector (mainly with specialised inputs) in complementing farmers' natural innovativeness.

For their part, while few farmers may possess, say, the skills for specialised plant breeding or the equipment for agronomic or livestock experimentation, they do recognise a Good Thing when they see it. Their recognition may be aided by a background of formal education and experience and a foreground of information systems such as a knowledgeable extension officer or other forms of media.

Knowledge about new technologies must also extend to knowledge about the returns from adoption, which in a risky world necessitates judgements about alternative possible outcomes of yields and profits. These, in turn, depend on unknown weather, pest and price variables. Full information on risk is rarely available for new technologies, simply because they are new. Consequently, farmers' perceptions of risk may dominate the adoption decision in the early years, particularly if the early years should happen to be unfavourable.

If farmers form their risk perceptions in a rational way, with the passage of enough time, their perceptions will tend to converge to the objective risks relevant to their environment (O'Mara 1983). But in the early years, farmers may have exaggerated perceptions of the risks involved, and those who tend to underestimate the risks will adopt first. Of course, the more profitable a new technology is, on average, the less likely are risk perceptions to impede its adoption.

In addition to differentials in access to knowledge, farmers may be confronted by important differentials in access to inputs and markets. The availability of improved varieties may be limited at first, until seed producing specialists or public agencies have had time to multiply the initial stock. During this period, some farmers may have preferential access, and particularly the larger and more progressive farmers who are often selected by extension agents for early adoption, so that they can provide a favourable demonstration effect.

If the new technology involves a substantial increase in aggregate production, or the expansion of a new crop, there may be initial lags in the development of transportation, processing and marketing channels, which will act to slow the rate of adoption, particularly for those farmers more removed from the hub of market activities.

Differential access to credit is frequently offered as an explanation of why large scale farmers tend to adopt before small scale farmers. However, von Pischke (1978), and Perrin and Winkelmann (1976), have argued that own and informal credit is probably much more readily available than is generally thought within rural areas, and it is only when the new technology is but marginally superior to existing technologies that the subsidies inherent in most public credit schemes become important in swinging the decision to adopt.

Finally, one temporary phenomenon often overlooked is the possibility that, as more and more farmers adopt a yield improving technology, the increase in aggregate output will act to depress the market price. This effect will be greater the more inelastic the demand curve, and it can lead to two consequences. Kislev and Shchori-Bachrach (1973) have held that, as the price declines, the initial adopters will tend to give up the new technology in the quest for even more efficient methods of production. This argument is based on the assumption that the initial adopters are more skilled, and hence have higher opportunity costs for their labour. With higher marginal costs, they are the first to find the new technology unprofitable as the price declines. Nevertheless, the technology may continue to be attractive to more tardy adopters who have lower opportunity costs.

On the other hand, if a new technology is clearly superior and acts substantially to reduce coasts per unit of output (as with the HYVs), the price decline need not impede the adoption of the improved technology, but it may lead to substantial adjustments in the amount of the crop grown by different types of farms, including a reduction in the total area grown. Scobie and Posada (1978), for example, showed that the introduction of HYV rice in Colombia in the mid-1960s led to rapid adoption, but a subsequent sharp decline in rice prices bore very heavily on small unpland farmers, particularly on those farming in agroclimatic conditions unsuitable for the HYVs.

In a dynamic setting with continuously growing demand due to growth in population and incomes, these price effects should be short-term phenomena. They will not be so, however, if technological change is also dynamic, with a succession of cost-reducing technologies. Witness, for example, the long-term price problems in American agriculture.

Left to their own devices, short-term barriers to adoption should disappear with the passage of time. The Norfolk four-course rotation successfully spread throughout much of England in the second half of the eighteenth century, long before the days of government intervention through agricultural extension or farm credit schemes. Ruttan (1977) has also observed that, while smaller scale farmers and tenants tended to lag behing large scale farmers in the early years following the introduction of HYVs, these lags typically disappeared within a few years. More dramatically, they disappeared almost overnight in areas such as the Indian Punjab when HYV wheat was introduced in the mid 1960s.

Speedy adoption is more important today. Not only is the additional food required, but development experts and policy makers often need to see rapid success to justify their efforts and positions. When project funding is involved from international development banks, it is also necessary to show a respectable rate of return on the capital invested. With discounting, only those production increases obtained in the early years of a project have much impact on the overall rate of return.

The ceiling level of adoption is possibly most determined by the agro-climatic suitability of the new technology. New technologies that depend on irrigation will clearly not be grown on farms that do not have access to water. On a more subtle level, small variations in topography, soils, altitude or rainfall can make the difference between whether a new technology is suitable or not. Perrin and Winkelmann (1976) give an example from Turkey where improved wheat varieties were rejected in one village even though neighbouring villages had adopted them. The problem was found to be the slightly higher elevation of the rejecting village, and frost problems precluded use of the new varieties.

This problem is particularly acute in the development of improved crop varieties, and underscores the need for increased emphasis on local research to modify and adapt newly reconstituted genetic material to local conditions.

Labour bottlenecks can also be a real barrier to adoption for some kinds of farmers, and in some regions. High yielding crop varieties not only add to total labour requirements, but they often exacerbate seasonal peaks in labour requirements. Peaks typically occur at planting, weeding and harvest times. Also, if the new varieties have a shorter growing season, and permit additional multiple cropping, there may be a consequent overlapping of the harvesting and planting of successive crops, with very sharp increases in seasonal labour requirements. Unless local labour markets are elastic, increases in seasonal wage rates quickly dampen the profitability of new technologies, particularly for larger farms which cannot get by with family labour alone. Adoption may then depend on complimentary and expensive investments in farm mechanization.

Protracted difficulties in obtaining fertiliser or pesticides may also be a barrier to adoption of some technologies. The problem is often not so much the general availability of these inputs but the difficulty of getting them at the right time. Supplies of fertiliser and pesticides may also be very uncertain, being subject to the vagaries of government policy in dealing with foreign exchange crises.

Risk aversion may also play a role in long-term adoption problems. Farmers are generally averse to risk (Anderson, Dillon and Hardaker 1977), although there is considerable variation in their behavioural patterns. If a new technology is clearly superior to an established technology in the sense that the return will be greater no matter what happens to the weather, prices, etc., risk aversion is not likely to deter farmers from adopting (Anderson 1974). But if the new technology is not superior under all possible eventualities, then differences in risk attitudes can play a role, even when perceptions about the riskiness of the new technology are correct.

Small scale farmers are usually thought to be more risk averse than larger scale farmers, because they can least afford to take risks. If so, they should be more reluctant to adopt new and risky technologies, even when these are more profitable on average. Surprisingly, the empirical evidence on the importance of risk in adoption decisions is not conclusive. Roumasset (1976) found that risk did not explain fertiliser decisions amongst Philippine farmers growing rice under irrigated conditions. Walker (1981) also found that adoption of a hybrid corn in El Salvador was not affected by differences in risk attitudes. On the other hand, there is considerable evidence to show that risk attitudes do affect cropping patterns (e.g., Hazell et al. 1983). Possibly, differences in the relative levels of risk involved in alternative crop technologies or production techniques are too small for risk to play an important role in these decisions. Risk is doubtless more important though in choosing amongst "lumpy" technologies, such as the purchase of livestock or tractors.

While there are a number of theoretical and empirical discussions about the possible impact of tenurial arrangements on adoption decisions, there is currently little consensus in the literature. In their review, Feder, Just and Zilberman (1982, p. 36) conclude that :

"... any observed effect of tenancy may be indirectly due to the implied relation between tenure and access to credit, input markets, product markets, and technical information. If these relationships differ in different sociocultural environments, empirical results may seem conflicting if the underlying factors are not considered directly. Thus, a lack of clear empirical results on the relationship between tenure and adoption may be due to the fact that many factors are yet to be considered appropriately".

Adoption studies often fail to consider the dynamics of the adoption process, and are undertaken too early to reveal the long-term barriers that determine the ceiling rate of adoption. Several studies fall in this category. They have lead to incorrect conclusions in the 1970s about the limitations of the HYVs. (See Farmer (1977) and Harriss (1977) ; HYVs are now almost universally grown in the North Arcot study area in south India).

Another problem has been the lack of an adequate theoretical model of adoption decisions to enable adequate distinction between true "barrier" variables and their surrogates. It seems likely, for example, that farm size is a surrogate for more fundamental problems such as access to credit and other inputs, capacity to bear risk, etc. (Feder, Just and Zilberman 1982). Thus, empirical verification that farm size affects adoption decisions is much less useful for policy prescription than identification of the underlying causes of the observed relation.

Adoption studies also need to recognise interdependencies between some of the restricting variables. Seasonal labour bottlenecks or risk considerations, for example, may on their own only explain a small part of the reason why farmers don't adopt. But when both factors are considered jointly, an overpowering case against adoption may sometimes emerge. These kinds of inter-relationships can be handled through multivariate regression analysis, where rates or levels of adoption are regressed against a set of explanatory variables. Another approach is to build mathematical programming models of the farmers' decision problems, and to attempt to simulate or predict their responses.

Benito (1976) constructed such a model for a peasant farm household in the Puebla area in Mexico. His model incorporated risk preferences, credit constraints, and the opportunity cost of labour in temporary and permanent non-farm employment. The new technology, in this case a package of inputs for increasing maize yields, also required that the farmer allocate time to organizational activities in the form of cooperative activities with other farmers to obtain the necessary inputs and credit. The model succesfully explained the low level of adoption observed in the Pueblo project after seven years of experience, and showed how farm size, risk, and off-farm job opportunities impinged on the adoption decision (see also Redclift 1983).

POLICIES FOR ACCELERATING TECHNOLOGICAL CHANGE IN AGRICULTURE

We began this essay with an expression of concern that, even though the worlds's agricultural growth rate has been impressive since the Second World War, it has not kept far enough ahead of population growth to

permit the possibility of widespread poverty alleviation. Of even greater
concern is the fact that the agricultural growth rate is slowing down.

The implications of these concerns are confounded by sharp differences
in individual country experiences. The USA and Western Europe, for example,
have enjoyed agricultural growth rates well in excess of the growth in their
own food needs. Their problem has been one of excess supply, strong downward forces on agricultural prices and farm incomes, and a need for public
support policies. In contrast, many developing countries have failed to meet
the growth in their food requirements because of rapid population growth
and, in some cases, because of sharp increases in per caput incomes (e.g.,
the OPEC countries).

These imbalances in food demand and supply can be, and to some extent are, resolved through international trade and food aid programs. A
problem remains though that the prices at which world markets clear lead
to inconsistencies between what poor people need and what they can afford.
Short of massive and sustained increases in food aid from the industrialised
countries, the only real solution for the poorer developing countries is to
increase sharply their own food production. As we have seen, the growth
linkage effects on incomes and employment from increased production also
create the possibility of simultaneously increasing the incomes of the poor,
thereby enabling them to purchase much of the additional food produced.

A major challenge for the developing countries in the years ahead is,
therefore, to accelerate their agricultural growth rates. The generation and
dissemination of new technologies will be the key to success. What policy
guidelines emerge from this review which may be useful in the task ahead?

Public Support for Research

The generation of modern agricultural technologies requires substantial
public support. The developing countries more than doubled their real aggregate expenditure on agricultural research between 1970 and 1980. The number of scientists also nearly doubled (IFPRI 1983). Yet despite these significant gains, real problems still exist in the distribution of research resources
between countries. About 80 per cent of the aggregate expenditure by developing counties on agricultural research is concentrated in about 20 relatively
large countries in Asia and Latin America. Many of the smaller and poorer
countries in Sub-Saharan Africa, Central America, and the Caribbean have
totally inadequate research systems. Given the long lags in agricultural research, high priority should be given by governments and donors to increasing the size and effectiveness of national research systems in the poorer coun-

tries. The International Agricultural Research Centres (IARCs) sponsored by the Consultative Group on International Agricultural Research (CGIAR) can play and are playing a critical role in assisting in this process of building up infrastructure, human capital and research resources in the Third World.

Enhancing the Effectiveness of Research

Research workers in agriculture have striven in many ways for greater relevance in their work. Traditionally, this relied heavily on the intuition and experience of people working closely with their clients. Over the middle decades of this century, however, the distance between clients (such as small scale farmers and landless labourers) and research personnel grew in several dimensions, such as the educational, linguistic, spatial, economic, social and bureaucratic, and renewed efforts are underway to reduce these. One recently popular approach to improving the feedbacks between innovation, research and application is farming systems research (FSR) variously reviewed and exposited by Shaner, Philipp and Schmehl (1982), Dillon and Anderson (1984) and Simmonds (1984). The nature and direction of rural research activities are explicitly influenced by formal perceptions of the wider environment in which identified clients find themselves (see Barker and Herdt 1982).

Of the ways of improving effectiveness of research, only a few others can be canvassed here. One general policy is through educational advancement, whether this be by lifting minimum educational achievements in communities at large, or by more targeted programs addressed, say, to improved skills in farming. In the latter, there is evidently much scope for identifying the targets more accurately and comprehensively to include, say, female members of the farming community and other oft-neglected groups such as remote resource-poor farmers and landless labourers.

Speeding the adoption of worthy improved technologies is another educational task that is usually tackled through a rural extension service, which variously works through personal contacts, written materials and electronic media. Bureaucracies being what they are, oportunities abound for improving the effectiveness of such services. Most obviously, their linkages to farmer clients and to research organisations need to be strengthened as a precondition to any degree of achievement (McInerney 1978, Jarrett, 1979). Budgetary support for a mobile, active, high quality, motivated extension service with good access to the media seems seldom to be very forthcoming, especially in the Third World. Until the situation is improved, the actuality of benefits from research and new technology will continue long to lag the potential.

Exploiting Genetic Resources

Reduction of the genetic diversity in plants, generally, and major food crops, in particular, is an important environmental risk associated with technological change. As the diversity decreases on farms, effective steps must be taken to ensure that the genetic material is maintained elsewhere. A considerable amount of work is under way in this area (Hawkes 1983, Chang 1984), although several issues remain to be resolved. These include : optimal interventions in the process of evolution which includes natural loss of many genetypes; optimal rates of collection, classification and storage/preservation of materials in the face of such loss; optimal rates of exploitation of "genebanks" of preserved germplasm in programs of breeding for improvement. Needless to say, uncertainty pervades the environment of all these optimal decisions so that they are not easy in any respect and, given the multidimensional character of, say, improvement (e.g., yield, stability, resistance, quality, etc.) are inherently complex.

Questing Knowledge for Improved Policy

The economics of the generation and exploitation of new technology, be it in agriculture, industry or elsewhere, is riddled with uncertainties. Uncertainties make the tasks of policy formulation and implementation very challenging (Ruttan 1977). Policy makers' lives would, in some senses, be easier if the uncertainties could be reduced through new knowledge from research. Research is well recognised as an inherently risky process and research on technology policy issues is no exception.

Some of the researchable issues of crucial relevance to agricultural development relate to alternative institutional models for R & D in developing counries. Some fundamental questions relate to the transferability of different technologies across agro-ecological, cultural and national boundaries (see Eicher 1984). To the extent that technologies are highly location-specific to be effective, the requisite R & D must be done more or less *in situ*.

Research in particular locations can be implemented with diverse degrees of external assistance. Recent decades have witnessed examples of alternatives across the range of possibilities, from indigenous (perhaps colonial) self help from local (including human) resources, through bilateral cooperative arrangements between local and more developed country institutions, to various multilateral arrangements involving several external donor agencies. One recent version of the latter has been the system of IARCs. There is no uniform operating procedure among the 13 centres comprising the system but, typically, they assemble a critical mass of research resources dealing with a hitherto neglected commodity or mandate region, and work collaboratively with national research systems.

The presumption has been that this latter model is a cost-effective mode of assistance. Our personal experiences and prejudices support the presumption, but it would be useful to donors and others to have more concrete evidence on the relative and absolute virtues of the various types of R & D assistance. Our expectation is that the relativities in particular would change as the national capacities for R & D evolve over time with experience and investment in human capital and infrastructure dealing with education, research and extension. Information on these future likely changes may presently be acquired by cross-country comparisons over a range of srages of development. We have such work in progress, although we do not underplay the difficulty of disentangling the confounding effects of cultural traditions, pricing policy, etc. from direct investment effects.

REFERENCES

AHLUWALIA, M.S. (1978), "Rural Poverty and Agricultural Performance in India", *Journal of Development Studies*, vol. 14(3), pp. 298-323.

——, (1985), "Rural Poverty, Agricultural Production and Prices : A Re-Examination", in J.W. Mellor and G.M. Desai (eds), *Agricultural Change and Rural Poverty : Variations on a Theme by Dharm Narain*, Johns Hopkins University Press, Baltimore, forthcoming.

AKINO, M. and HAYAMI, Y. (1975), "Efficiency and Equity in Public Research : Rice Breeding in Japan's Economic Development", *American Journal of Agricultural Economics*, vol. 57(1), pp. 1-10.

ANDERSON, J.R. (1974), "Risk Efficiency in the Interpretation of Agricultural Production Research", *Review of Marketing and Agricultural Economics*, vol. 42 (3), pp. 131-84.

——, DILLON, J.L. and HARDAKER J.B. (1977), *Agricultural Decision Analysis*, Iowa State University Press, Ames.

—— and HARDAKER, J.B. (1979), "Economic Analysis in the Design of New Technologies for Small Farmers", in A. Valdés, G.M. Scobie, and J.L. Dillon (eds), *Economics and the Design of Small Farmer Technology*, Iowa State University Press, Ames, pp. 11-26.

BARKER, R. and HERDT, R.W. (1982), "Setting Priorities for Rice Research in Asia", in R.S. Anderson, P.R. Bross, E. Levy znd B.M. Morrison (eds),

Science, Politics and the Agricultural Revolution in Asia, Westview for AAAS Boulder, pp. 427-61.

BELL, C., HAZELL, P. and SLADE, R. (1982), *Project Evaluation in Regional Perspective*, Johns Hopkins University Press, Baltimore.

BENITO, C.A. (1976), "Peasants' Response to Modernization Projects in Mini Fundia Economies", *American Journal of Agricultural Economics*, vol. 58 (2), pp. 143-51.

BINSWANGER, H.P. and RUTTAN, V.W. (eds) (1978), *Induced Innovation : Technology, Institutions and Development*, Johns Hopkins University Press, Baltimore.

BLYN, G. (1983),"The Green Revolution Revisited", *Economic Development and Cultural Change*, vol. 32 (3), pp. 705-725.

BOSE, S.R. (1982), "Has Economic Growth Immiserized the Rural Poor in Indonesia? A Review of Conflicting Evidence", In G.B. Hainsworth (ed), *Village-Level Modernization in Southeast Asia*, University of British Columbia Press, Vancouver, pp. 53-69.

BYERLEE, D. and HARRINGTON, L. (1983), "New Wheat Varieties and the Small Farmer", in B.L. Greenshield and M.A. Bellamy (eds.), *Rural Development : Growth and Equity*, Gower for International Association of Agricultural Economists, Aldershot, pp. 87-92.

CAPALBO, S.M. and TRANG, T.V. (1984), "A Selected Survey of Recent Econometric Evidence Regarding Productivity and the Structure of U.S. Agriculture". Paper prepared for the Workshop on Developing a Framework for Assessing Future Changes in Agricultural Productivity, sponsored by the National Center for Food and Agricultural Policy, Resources for the Future, Washington, D.C., (July 16-18).

CHANG, T.T. (1984), "Conservation of Rice Genetic Resources : Luxury or Necessity?", *Science*, No. 224 (4646), pp 251-6.

CHAUDHRY, M.G. (1982), "Green Revolution and Redistribution of Rural Incomes : Pakistan's Experience", *Pakistan Development Review*, vol. 21 (3), pp. 173-205.

CHAYANOV, A.V. (1925), *The Theory of Peasant Economy*, translated by D. Thorner, B. Kerblay and R.E.F. Smith (1966), Irwin, Homewood.

CIAT (1981), *Report on the Fourth IRTP Conference for Latin America*, Cali, August.

CIMMYT (1978), *CIMMYT Review*, El Batan, Mexico.

CLAYTON, E. (1983), "Agricultural Development and Farm Income Distribution in LDCs", *American Journal of Agricultural Economics*, vol. 34(3), pp. 349-59.

COLE, K., CAMERON, J. and EDWARDS, C. (1983), *Why Economists Disagree*, Longman, London.

DEANE, P. (1965), *The First Industrial Revolution*, Cambridge University Press.

DE JANVRY, A. (1975), "The Political Economy of Rural Development in Latin America: An Interpretation", *American Journal of Agricultural Economics*, vol. 57 (3), pp. 490-9.

——, (1977), "Inducement of Technological and Institutional Innovations: An Interpretative Framework", in T.M. Arndt, D.G. Dalrymple and V.W. Ruttan (eds), *Resource Allocation and Productivity in National and International Agricultural Researsch*, University of Minnesota Press, Minneapolis, pp. 551-63.

DILLON, J.L. and ANDERSON, J.R. (1984), "Concept and Practice of Farming Systems Research", in *Proceedings of the Eastern Africa — ACIAR Consultation on Agricultural Research, 18-22 July 1983, Nairobi*, Australian Centre for International Agricultural Research, Canberra, pp. 171-86.

EICHER, C.K. (1984), "International Technology Transfer and the African Farmer: Theory and Practice", Working Paper No. 3/84, Dept. of Land Mgt, Univ. of Zimbabwe.

EMMANUEL, A. (1969), *L'Echange Inegal*, Maspero, Paris.

EVENSON, R.E. and FLORES, P.M. (1978), "Social Returns to Rice Research", in IRRI, *Economic Consequences of the New Rice Technology*, Los Banos, Philippines.

FARMER, B.H. (ed), (1977), *Green Revolution?*, Macmillan, London.

FEDER, G., JUST, R.E. and ZILBERMAN D. (1982), *Adoption of Agricultural Innovations in Developing Countries: A Survey*, World Bank Staff Working Paper No. 542.

FRAENKEL, F.R. (1976), *India's Green Revolution: Economic Gains and Political Costs*, Princeton University Press.

GIBB, A. Jr. (1974), "Agricultural Modernization, Non-Farm Employment and Low-Level Urbanization: A Case Study of a Central Luzon Sub-Region". Unpublished Ph. D. thesis, University of Michigan.

GRIFFIN, K. (1972), *The Green Revolution: An Economic Analysis*, UNRISD, Geneva.

—— (1979), *The Political Economy of Agrarian Change*, 2nd edn, Macmillan Press, London.

—— and GHOSE, A.K. (1979), "Growth and Impoverishment in the Rural Areas of Asia", *World Development*, vol. 7 (4/5), pp. 361-83.

GRIGG, D.B. (1984), "The Agricultural Revolution in Western Europe", in T. Bayliss-Smith and S. Wanmali (eds), *Understanding Green Revolutions*, Cambridge University Press, pp. 1-15.

GURLEY, J.G. (1975), *Challenges to Capitalism . Marx Lenin, and Mao*, Stanford Alumni Association.

HARADKER, J.B., ANDERSON, J.R. and DILLON, J.L. (1984), "Perspectives on Assessing the Impacts of Improved Agricultural Technologies in Developing Countries", *Australian Journal of Agricultural Economics* vol. 23 (2 and 3), pp. 87-108.

HARRISS, J. (1977), "The Limitations of HYV Technology in North Arcot District : The View from a Village" in B.H. Farmer (ed.), *Green Revolution*, Cambridge University Press.

HAWKES, J.G. (1983), *The Diversity of Crop Plants*, Harvard University Press, Cambridge.

HAYAMI, Y. and RUTTAN, V.W. (1971), *Agricultural Development : An International Perspective*, Johns Hopkins Unversity Press, Baltimore.

HAZELL, P.B.R. (1984), "Sources of Increased Instability in Indian and U.S. Cereal Production", *American Journal of Agricultural Economics*, vol. 66 (3), pp. 302-311.

——, NORTON, R.D., PARTHASARATHY, M. and POMAREDA, C. (1983), "The Importance of Risk in Agricultural Planning Models", in R.D. Norton and L. Solis M. (eds), *The Book of CHAC : Programming Studies for Mexican Agriculture*, Johns Hopkins University Press, Baltimore, pp. 225-49.

HERDT, R.W., and CAPULE, C. (1983), *Adoption, Spread, and Production Impact of Modern Rice Varieties in Asia*. IRRI, Los Banos, Philippines.

HEWITT DE ALCANTARA, C. (1976), *Modernizing Mexican Agriculture*, UNRISD, Geneva.

HIRSCHMAN, A.O. (1959), *The Strategy of Economic Development*, Yale University Press, New Haven.

IFPRI (1983), *IFPRI Report*, Washington, D.C.

ILO (1977), *Poverty and Landlessness in Rural Asia*, ILO, Geneva.

JAMES, C. (1983), "Wheat and Maize : CIMMYT'S Experience", *The Courier*, No. 82, pp. 63-65, (November-December).

JARRETT, F.G. (1979), "Technological Change-Creator or Destroyer", in A.T.A. Healy (ed.), *Science and Technology for What Purpose? An Australian Perspective*, AGPS, Canberra, pp. 109-24.

JOHNSTON, B.F. and KILBY, P. (1975), *Agriculture and Structural Transformation*, Oxford University Press, New York.

KIKUCHI, M. and HAYAMI, Y., "Technological and Institutional Response and Income Shares under Demographic Pressure : A Comparison of Indonesian and Philippine Villages", in G.B. Hainsworth (ed.), Village-

Level Modernization in Southeast Asia, University of British Columbia Press, Vancouver, pp. 173-90.

KISLEV, Y. and SHCHORI-BACHRACH, N. (1973), "The Process of an Innovation Cycle", *American Journal of Agricultural Economics* vol. 55 (1), pp. 28-37.

LAPPÉ, F.M. and COLLINS, J. (1980), *Food First : Beyond the Myth of Scarcity*, Ballantine, New York.

LELE, U. and MELLOR, J.W. (1981), "Technological Change, Distribution Bias of Labour Transfer in a Two Sector Economy", *Oxford Economic Papers*, vol. 33(3), pp. 426-41.

LENIN, V.I. (1899), "The differentiation of the Peasantry", Reprinted from his *Collected Works*, Lawrence and Wishart, London, 1960 in J. Harriss (ed.) (1982), *Rural Development : Theories of Peasant Economy and Agrarian Change*, Hutchinson, London.

LIPTON, M. (1977), *Why Poor People Stay Poor : Urban Bias in World Development*, Harvard University Press, Cambridge.

—— (1978), "Inter-farm, Inter -regional and Farm-non-farm Income Distribution: The Impact of the New Cereal Varieties", *World Development*, vol. 6 (3), pp. 319-37.

McINERNEY, J.P. (1978), *The Technology of Rural Development*, World Bank Staff Working Paper No. 295, Washington, D.C.

MELLOR, J.W. (1973), "Accelerated Growth in Agricultural Production and the Intersectoral Transfer of Resources", *Economic Development of Cultural Change*, vol. 22(11), pp. 1-16.

—— (1975), *The Impact of New Agricultural Technology on Employment and Income Distribution - Concepts and Policy*, US. Agency for International Development, Occasional Paper No. 2, Washington, D.C. (May).

—— (1976), *The New Economics of Growth*, Cornell University Press, Ithaca.

—— and LELE, U. (1973), "Growth Linkages of the New Foodgrains Technologies", *Indian Journal of Agricultural Economics*, vol. 26 (1), pp. 35-55.

O'MARA, G. (1983), "The Microeconomics of Technique Adoption by Smallholding Mexican Farmers" in R.D. Morton and L. Solis, (eds.). *The Book of CHAC: Programming Studies for Mexican Agriculture*, Johns Hopkins University Press, Baltimore, pp. 250-89.

PAULINO, L. (1984), "Global Trends in Cereal Supply and Demand and Their Implications to Price Environment". Paper prepared for the IFPRI Workshop on Food and Agricultural Policy, Belmont Estate, Elkridge, Maryland, (April 29-May 2).

PERRIN, R. and WINKELMANN, D. (1976), "Impediments to Technical Progress on Small Versus Large Farms" *American Journal of Agricultural Economics*, vol. 58 (5), pp. 888-94.

PINSTRUP-ANDERSEN, P. (1979), "The Market Price Effect and the Distribution of Economic Benefits from New Technology", *European Review of Agricultural Economics*, vol. 6 (1), pp. 17-46.

—— (1982), *Agricultural Research and Technology in Economic Development*, Longman, London.

—— and HAZELL, P.B.R. (1984), "The Impact of the Green Revolution and Prospects for the Future", in I. Hornstein and R. Terarishi (eds), *Food Review International*, vol. 1, Marcel Dekker, New York, (in press).

PRAHLADACHAR, M. (1983), "Income Distribution Effects of the Green Revolution in India : A Review of the Empirical Evidence", *World Development*, vol. 11(11), pp. 927-44.

PREBISCH, R. (1959), "Commercial Policy in Underdeveloped Countries", *American Economic Review*, vol. 44 (2), pp. 251-73.

QUIZON, J.B. and BINSWANGER, H.P. (1983), "Income Distribution in Agriculture : A Unified Approach", *American Journal of Agricultural Economics*, vol. 65(3), pp. 526-38.

RANGARAJAN, C. (1982), *Agricultural Growth and Industrial Performance in India*, IFPRI Research Report No. 33, Washington, D.C.

REDCLIFT, M. (1983), "Production Programs for Small Farmers: Plan Puebla as Myth and Reality", *Economic Development and Cultural Change*, vol. 31(3), pp. 551-70.

RESOURCES FOR THE FUTURE (1984), "Feeding a Hungry Wolrd", *Resources*, No. 76 (Spring), pp. 1-20.

ROUMASSET, J.A. (1976), *Rice and Risk : Decision Making among Low-Income Farmers*, North-Holland, Amsterdam.

RUTTAN, V.W. (1956), "The Contribution of Technological Progress to Farm Output : 1950-75", *Review of Economics and Statistics*, vol. 38 (2), pp. 61-69.

—— (1960), "Research on the Economics of Technological Change in American Agriculture", *Journal of Farm Economics*, vol. 42(4), pp. 735-54.

—— (1977), "The Green Revolution : Seven Generalizations", *International Development Review*, vol. 4, pp. 16-27.

—— (1982), *Agricultural Research Policy*, University of Minnesota Press, Minneapolis.

SARMA, J.S. (1982), *Agricultural Policy in India : Growth with Equity*, Report No. 201-e, IDRC, Ottawa.

SCOBIE, G.M. (1979), *Investment in International Agricultural Research : Some Economic Dimensions*, World Bank Staff Working Paper No. 361.

—— and POSADA R. (1978), "The Impact of Technical Change on Income Distribution : The Case of Rice in Colombia", *American Journal of Agricultural Economics*, vol. 60 (1), pp. 85-92.

SHANER, W.W., PHILIPP, P.F. and SCHMEHL, W.R. (1982), *Farming Systems Research and Development : Guidelines for Developing Countries*, Westview, Boulder.

SIMMONDS, N.W. (1984), *The State of the Art of Farming Systems Research*, World Bank, Washington, D.C., (in press).

STEVENS, R.D. (1977), "Transformation of Traditional Agriculture : Theory and Empirical Findings", in R.D. Stevens (ed.), *Tradition and Dynamics in Small-Farm Agriculture*, Iowa State University Press, Ames, pp. 3-24.

SWEEZY, P.M. (1968), *The Theory of Capitalistic Development*, Modern Reader, New York.

VON PISCHKE, J.D. (1978), "When is Smallholder Credit Necessary?, *Development Digest*, vol. 16 (3), pp. 6-14.

WALKER, T.S. (1981), "Risk and Adoption of Hybrid Maize in El Salvador", *Food Research Institute Studies*, vol. 28 (1), pp. 59-88.

WARREN, G.F. and LIVERMORE, K.C. (1911), *An Agricultural Survey*, Bulletin No. 295, Dept. of Farm Management, Cornell University, Ithaca, pp. 377-569.

CHAPTER NINE

INFORMATION AND TECHNOLOGICAL CHANGE — A RESEARCH PROGRAM IN RETROSPECT

By D. McL. Lamberton*, Stuart Macdonald*, and Thomas D. Mandeville*

THE INFORMATION RESEARCH UNIT (IRU) — ITS FIRST FIVE YEARS

This essay takes up an unusual invitation : to review our own recent work. Since the IRU's beginning within the Department of Economics at the University of Queensland in 1979, some research funding has come in and a lot of publications have gone out[1]. But just what has really been achieved?

Some scholars in the industrial era refused to summarise their work. Their rationale was that since so much effort had gone into each publication to explain the intricacies in each instance, it would be futile to gloss over all the detail that justified and gave meaning, in order to provide the lazy with a superficial summary. Alas, we live in more pragmatic, information-over-loaded times. Rarely will anyone these days bother to plough through detail to determine its significance. This task, then, falls on the knowledge producer with inevitable bias about the relative value of the work.

The work of the IRU has been research (Table 1); the founding, editing and publishing of a new journal, *Prometheus*; as well as various other teaching, editing, and communicating activities[2]. Prior to the advent of *Prometheus*[3] in June 1983, no single journal dealt exclusively with the areas of technological change, innovation, information economics, communications, and science policy. Now into its third issue[4], and almost breaking even on costs, *Prometheus* has begun to establish itself as a forum for considered debate on these issues.

The ill-fated Myers Inquiry into Technological Change in Australia was the catalyst that brought the authors — the founding members of the IRU — together as a working group in mid-1979. Previous work by individual mem-

---

* University of Queensland, Australia.

1. The reference list includes selected items only.

2. Editorial activities include involvement in *Information Economics and Policy,* as well as the recent collection of original essays combined in *The Trouble with Technology* (Macdonald, Lamberton, Mandeville, 1983).

3. *Prometheus*, Vol. 1.

4. *Prometheus*, Vol. 2.

bers provided a foundation for the new Unit's activities[5]. For the Myers Inquiry, studies of the diffusion and employment effects of computers in small business (Macdonald, Mandeville and Lamberton, 1980) and local government (Mandeville and Macdonald, 1981), were undertaken along with

TABLE 1

IRU Research Projects

(Classification of Selected Items in the Reference List by Subtopics)

I.   Diffusion and Employment Effects of Information Technology
37, 45, 25. 33, 2, 47.

II.   "Xerox Effect" Hypothesis
26

III.   Real Costs of Research in Universities
39

IV.   The Patent System, Individual Inventors, and Information Sources
43, 29, 40.

V.   Telecommunications
      A.   International flows
          36, 42, 27.
      B.   Deregulation
          38
      C.   Role in LDCs
          3, 15, 16, 19, 4.
      D.   Tradeoffs
          41

VI.   Information Sector
12, 15, 47, 5, 6.

VII.   Innovation in Agricultural History
28, 32.

VIII. Information and Economic Theory
7, 8, 17, 21, 22, 23.

IX.   Critical Assessment of Science, Technology, and Information Policy in Australia
44, 46, 13, 31, 30, 43, 42, 14, 18, 20, 24.

---

a study of the word processor industry in Australia (Lamberton, Macdonald and Mandeville, 1980). The Myers Inquiry saw fit to publish only the latter study. Perhaps this was because our conclusions in the computer and small

5. For example, Lamberton (1971, 1974, 1975, 1976, 1975-1977, 1977), Braun and Macdonald (1978), Macdonald (1975).

business study were considerably less sanguine about employment effects than was the thrust of the Myers Report.

Further applied work on the employment effects of information technology in the banking (Macdonald, Lamberton and Hodge, 1981) and insurance (Brown and Macdonald, 1982) industries, (funded respectively by the Reserve Bank of Australia and the University of Queensland) as well as theoretical work on the role of information in the economy funded by the Australian Research Grants Scheme, eventually led to the formulation of our "Xerox effect" hypothesis — that information technology may actually be eroding productivity levels (Lamberton, Macdonald and Mandeville, 1982).

Early in 1980 the University of Queensland commissioned the IRU to investigate the costs of research in that university. It was found that the bulk of research costs were indirect, and one of the implications of this was that the real costs of allocating internal research monies often greatly exceeded the value of research grants (Macdonald, Mandeville and Lamberton, 1982).

Also in 1980, in connection with the government Inquiry into the Australian Patent System, the Australian Government commissioned the IRU to conduct an economic assessment of the Australian patent system (Mandeville, Lamberton and Bishop, 1982) as well as a study of the role of the individual inventor (Macdonald, 1983a). These projects involved extensive surveys of industry, patentees, engineers, and individual inventors. As a by-product, these studies shed considerable light on the sources of technological information in Australia and emphasised personal contact as the prime source (Mandeville, 1983a).

By 1981, while the patent studies were still in progress, work had begun on resource allocation and public accountability in the government research organisation CSIRO (Macdonald, 1983b) as well as on case studies of innovation in Australian industry. Over the 1981-83 period the group did considerable work in the telecommunications area. The ASEAN-Australia Joint Research Project commissioned a study to assemble basic data on the economic feature of the telecommunications industry within these countries, as well as the extent of telecommunications traffic among them (Macdonald and Mandeville, 1984). An internal research grant from the University of Queensland enabled work to begin on analysing information flows between the Australian and Japanese economies (Mandeville, forthcoming). Case studies of transborder data flows and large Australian firms were funded jointly by the OECD and the Department of Communications (Lamberton and Mandeville, 1983). Work was also undertaken on deregulation as well as the ability of telecommunications authorities to respond to technological change and increasing competition from the private sector (Macdonald, Mandeville and

Lamberton, 1981), the role of telecommunications in LDCs (Jussawalla and Lamberton, 1982; Lamberton, 1982b, 1982c, 1983c; Karunaratne, 1982), and possible trade-offs between telecommunications and transportation (Mandeville, 1983b).

In 1983 the Queensland Department of Employment and Labour Relations commissioned a study of technological change and the information sector in the State of Queensland - one of the first regional information sector studies conducted in the Machlup-Porat tradition (Mandeville, Macdonald, Thompson and Lamberton, 1983). Also since 1983, collaboration between the East-West Communication Institute (Honolulu, Hawaii) and the IRU has extended research on national information sectors to small and developing economies of the Pacific Region. Information sector measurement has been carried out for Australia, Fiji, Papua New Guinea, New Zealand, Singapore, Philippines and Indonesia (Karunaratne and Cameron, 1980, 1981).

During the current year (1984) the OECD is supporting work on the issue of protection and innovation, while an internal University grant is providing funds for the investigation of the self-serviced household economy. Work is also in progress on a study of the technology intensity of trade, a study of state technology policies, further information sector studies in collaboration with the East-West Communication Institute, and a study of the productivity of new office technology. In 1985 it is expected that the Unit will undertake a major economic evaluation of a new medical technology as well as a study of the economics of cable television. By way of contrast to the current nature of much of the IRU's work, a study of the role of the agricultural labourer in innovation in nineteenth century Britain has just been completed (Macdonald, 1983d).

Members of the IRU have directly contributed to various aspects of the technology debate in Australia. Over the 1978-81 period, this debate was primarily concerned with the social and economic effects of technological change. Critical assessments were made of the quality of the debate (Mandeville and Macdonald, 1980), the employment issue (Macdonald and Mandeville, 1980), the difficulties of modelling the future impact of technological change (Mandeville, Macdonald and Lamberton, 1980) and aspects of social costs (Lamberton, 1981).

For the past two years or so the national technology debate has focussed on Australia's appalling innovation performance and possible high technology opportunities, as well as the role of CSIRO, the patent system, Australia's changing relationship with Japan and national information policy. IRU members have criticised the amateurish approach to high technology policy (Macdonald, 1983c), demonstrated and criticised the lack of public accountability of CSIRO (Macdonald, 1983b), shown that the costs of the patent system probably exceed its benefits (Mandeville, Lamberton and Bishop, 1982), pointed to weak technology links with Japan (Mandeville, forthcoming), and urged the need for a national information policy (Lamberton, 1982a, 1983b, 1984a, forthcoming c).

Finally, the role of information in economic theory and analysis has been of long standing and continuing interest (Lamberton, 1971, 1974, 1983a, 1984b, forthcoming a, forthcoming b). Indeed, as will be made apparent in the final section of this paper, the focal point of the IRU's work is information. This is best perceived against the background development of information economics itself.

INFORMATION ECONOMICS

Information economics[6] might appear to be of recent origin because it was officially recognised by the American Economic Association in 1976 by award of a category in the Association's classification System for Articles and Abstracts. Such an interpretation of its history fits the popular but erroneous view that information economics is simply a reflection of the spectacular advent of intelligent electronics with its greatly enhanced capacities for communication, computation, and control. The view is erroneous because all societies have been information societies and have employed information technologies. What has been changing in response not only to computers and satellites but also to recognition of the deficiencies of economic theory and failures of government and business policies, is the role assigned to information in economic analysis. Greater concern with the present and its problems rather than long-run equilibrium has led to information activities being seen as cause of disequilibrium and means to equilibrium; as endogenous rather than exogenous. This shift of emphasis has been most obvious in the case of technology, which is perhaps the most important and potentially beneficial kind of information.

A decade ago those interested in information matters still emphasised the need to incorporate a role for information in economic models because of its bearing upon market performance. Pioneers like Jacob Marschak and Fritz Machlup had already done much to shape the pattern of development of information economics. Marschak had initiated true theoretical work and established a link with the study of organisation as an information-handling

6. A recent review of the literature in information economics is contained in Lamberton (1983a).

decision system. Machlup had provided a detailed statistical account of information activities. The radical thought that is only now emerging with increasing clarity is that organisational change as well as technological change was fundamentally important and could be analysed in economic terms.

Information and the information-handling mechanisms we call organisations were now to be treated as resources. But it seems that generations will have passed before the full implications are understood. Put in the simplest terms possible, organisation is now to be a variable, as the product was made a variable in the exciting days of imperfect and monopolistic competition theory. This provides a link between the Marschak and Machlup contributions. The decision theory framework of Marschak gives a role to information which, depending upon the costs of information, shape the pattern of information activities recorded in such rich detail by Machlup and his followers.

To put these thoughts in a more technical fashion, the probability distribution of states of the world is treated as a variable. The decision-maker can take advantage of the existence of signals conveying messages about the environment and can choose which signals to acquire. There can be great advantages in co-operating with others in information-handling, i.e., forming an organisation. This calls for creation of information channels, the building up of a stock of information, and the creation of an "organisation" language that may limit access to non-members. The technical characteristics of these resources have two major implications : first, random events can have considerable importance; and, secondly, the more successful the persuit of efficiency, the more rigid and unresponsive the organisation may become. This line of thought leads to an understanding that decision-makers may become locked into their information systems just as easily as into stocks of building and machines. Organisational sclerosis is in this way given an economic rationale.

The subject matter of information economics is, of course, neither more nor less than that of economics itself, because all decision-makers, except those in microeconomics textbooks are involved in choosing which information to acquire and use. Even rational expectations theory must recognise that there is a wide range of procedures that economic actors use to cope with uncertainty. When circumstances dictate that reliance upon these procedures are important, as in dynamic conditions, the theorist is once more in need of a theory of the decision-making process.

It is, however, possible to single out some of the important economic characteristics of information and to emphasise some propositions central to information economics :

1. The difference between personal and group or organisational use of information. As Kenneth Arrow has argued, the division of information gathering is perhaps the most fundamental form of the division of labour.

2. The cost of producing information is independent of the scale on which it is used.

3. The greater part of the cost of information is the cost incurred by the recipient.

4. The stock of information and the organisations created to handle information have the characteristics of capital.

5. The complexity of information activities makes information as a resource difficult to contain within the traditional production function analysis.

6. The limitations on information as a commodity dictate resort to organisation as an alternative to markets.

Research, both theoretical and empirical, continues on a wide front and publication regularly reaches to every category in the entire American Economic Association classification. The activities catalogued in the first section of this paper show this diversity and, we hope, illustrate some or all of the above mentioned characteristics and propositions. In the third and final section an attempt is made to identify the central theme of IRU research.

THE FOCAL POINT OF IRU RESEARCH

Drawing together the research work of several people over several years so that that work may be seen to have a single point, in sharp focus, is unlikely to be an easy exercise, even when nearly all the research has been collaborative. Circumstances change — especially the constraints on this sort of research — interest and opportunities alter, some past mistakes are recognised and their repetition is avoided. The direction of research should change over time; the alternatives are stagnation and irrelevance. In a research area itself concerned with change and its agents, change in the research is mandatory. That is not to say that a theme and a style and an approach cannot be maintained. They can, and they have in the work of the IRU. Indeed, they have been essential in distinguishing the integrity of the Unit's research. There has been no ready-made niche within *conventional* economics in which this research could comfortably claim status and recognition — nor within any other discipline for that matter. The research work has had to stand on its own, largely un-buttressed and non fortified by established doctrine, and has had to depend on its own integrity for support. Indeed, the research work has had to withstand some battering from established disciplines, institutions and organisations in that it has questioned their behaviour and rationale. A cynic might even suggest that the questioning of established thinking was the theme — the focus — of IRU research. That would be only indirectly correct: the research focusses on information — its qualities, its use, its production, its transfer and its impact. Research of such a fundamental nature inevitably beings conflict with those who have long assumed that the role of information is obvious and the problems it poses inconsequential.

In its concern with information in the broadest context, the IRU has had little to do with research on the hardware — the machinery — that is commonly seen to be responsible for the "Information Revolution". Many others work on such matters, but that is not the main reason for our virtual abstention. The problems and opportunities of information extend far beyond the means by which it may be collected, stored and processed. Indeed, producing information has now become a relatively cheap and easy task compared with the horrendous — and largely unrecognised — problem of using information efficiently and equitably. The latter is a much more serious and intellectually challenging issue and has consumed much of the Unit's research effort. That is not to say that machinery has been neglected entirely; that would be a cavalier and unrealistic attitude. It has been considered carefully in studies of the diffusion of word processors and of computers in small business, in a study of growing deregulation in telecommunications, in research on new technology in the banking and insurance industries, and necessarily in minute detail in research on the development of the electronics industry. The role of hardware must be fully appreciated before other matters can be adequately explored; ignorance of that role only tempts fundamental errors in further exploration. The emphasis of IRU research work has always been on the area beyond the hardware, the area in which personal and organisational behaviour in the use of information are of paramount importance. Thus, for example, we have found that the adoption of new information machinery by the banks has no direct influence on employment in those banks. What does is the way the banks choose, or are able, to use new machinery. Organisational constraints determine how the machinery is used and it is through the organisation, rather than directly, that such machinery exerts an impact on employment.

Perhaps the single greatest error in the work of the IRU has been a misconception rather than pure error. Much early work was concerned in the fashion of the times, with the impact of new technology on employment.

That research was well worth performing, if only for the learning experience gained. That experience has shown that it is unhelpful to regard employment as suffering an impact from technology. This is not just because the influence of new technology is determined by the nature of the adopting organisation; more important is the appreciation that many of those factors — such as employment — which are commonly regarded as being affected by technological change are, in fact, an integral part of that very process of technological change. Experience has confirmed that the process does not start with invention and caese with innovation as the traditional linear model would have it. The whole process of technological change — its creation, introduction and diffusion — is a single innovative process, an information process to which information from employees or customers makes as much contribution as information from scientists and engineers. It is quite artificial to construe new technology as the product of research and development which goes forth to exert an impact on the world beyond. Technology is largely information (even though some of that information may be embodied in a machine), and technological change is an information process which continues well beyond innovation as further bits of information are contributed from diverse and distant peripheral sources — as they were from core research — to create a total information package. It is not that employment considerations are irrelevant to the process of technological change; far from it — they are an integral part of that process.

Some of the Unit's most intriguing research has been concerned with the flow of information — where it comes from, where it goes to and the means by which it travels. Included in this research is work on the sort of information produced and the ability of others to use this information. In Australia, where nearly 80 per cent of research and development is performed in the public sector, such work has obvious science policy implications. It would seem that the private sector receives very little of the information generated by public sector R & D. Indeed, those institutions which are supposed to function as information disseminators (such as the patent system), seem to be responsible for little of the total information used by the world outside the institution, and for none that is of use in isolation from other information. Information often travels best embodied in people; personal and informal contact seems to be essential to effect information transfer. The more information-intensive the activity — and high technology is an excellent example of such an activity — the more important such contact becomes. This has direct relevance for high technology policy, and especially for the many efforts currently underway in Australia and elsewhere to stimulate high technology through institutional means. A landscaped high technology park at the gates of a university is not the clear guarantee of universal employment, wealth generation and of a costlessly restructured economy that so many fondly imagine.

The Unit's research is radical and distinctive because it regards information as an essential factor in all the research areas it covers — from agricultural improvement in nineteenth century England to manpower planning in modern Australia. We believe this approach to be a valuable alternative to traditional methodology. In its most evident form it challenges the usefulness of a three-sector model of the economy which disregards the single most significant wealth-creating factor. Understanding of a modern economy cannot but be deficient without an appreciation of the information sector. A less obvious example is an examination of Australia's largest national research laboratories, not in terms of what they do, but in terms of the information available about what they do. That approach presents the more traditional issue in a new light altogether. Similarly, the Unit's studies of the real costs of university research have concentrated on the total costs of creating research information and on the costs of providing information by which research proposals may be assessed. Both costs are much higher than a conventional analysis would have suggested. Similarly, it is possible — indeed, necessary — to examine certain techniques in terms of the broad information constraints on the use of those techniques. It is not, for example, easy to employ input-output techniques to forecast the impact of technological change. An Australian attempt to do so clearly illustrated the difference between the broad body of received information that is technology, and the mere tool which is technique. Technique is a dangerous weapon in the hands of those without technology.

There is always a danger that the research work of any individual, or any individual group, will become introverted and isolated from the work of others. That danger is especially real when the area of research is unusual. There is a further risk that the pressure to produce yet more research results to justify the Unit's approach may produce quantity at the expense of quality. Hopefully, recognition of the dangers will help allay them. So too should the Unit's editing of both an international collection of original papers in the field, and of the new international journal, *Prometheus*. There are, though, obstacles which are more difficult to overcome. The Unit's published research has endeared it neither to those who value only a conventional approach and are unimpressed by the importance of information, nor to those who are in a position to fund research.

Probably the former will have to be converted before the latter will become allies. Until then, the Information Research Unit will continue its efforts as best it can and until then it will remain poor but honest.

REFERENCES

1. BRAUN, E. and MACDONALD, S. (1978), *Revolution in Miniature - History and Impact of Semiconductor Electronics*, Cambridge University Press.
2. BROWN, A. and MACDONALD, S. (1982), "Technological Change and Employment in the Insurance Industry", *Economic Activity*, vol. 25.
3. JUSSAWALLA, M. and LAMBERTON, D.M. (eds) (1982), *Communication Economics and Development*, Pergamon.
4. KARUNARATNE, N.D. (1982), "Telecommunication and Information in Development Planning Strategy", in M. Jussawalla and D.M. Lamberton (eds), *Communication Economics and Development*, Pergamon.
5. —— and CAMERON, A.D. (1980), "Input-Output Analysis of the Australian Information Economy", *Information and Management*, vol. 5.
6. —— (1981), "A Comparative Analysis of the 'Information Economy' in Developed and Developing Countries", *Journal of Information Science*, vol. 3.
7. LAMBERTON, D.M. (ed.), (1971), *Economics of Information and Knowledge*, Penguin.
8. —— (ed.), (1974), *The Information Revolution*, The Annals of the American Academy of Political and Social Science, vol. 412.
9. —— (1975), *Who Owns the Unexpected?*, University of Queensland Press.
10. —— (co-author) (1976), *Public Libraries in Australia*, Report of the Committee of Inquiry into Public Libraries, Australian Government Publishing Services.
11. —— (co-author), (1975-1977), *The Australian Information Industry*, vols, 1-12, Reports to Telecommunications Planning Unit, Telecom Australia, University of Queensland.
12. —— (1977), "Structure and Growth of the Communications Industry", in K.A. Tucker (ed.), *Economics of the Australian Service Sector*, Croom Helm.
13. —— (1981), "Social Costs of Technological Change", in OECD *Information Activities, Electronics and Telecommunications Technologies : Impact on Employment and Growth, vol. 2*, OECD.
14. —— (1982a), *Policies and Programmes in Stimulation of Innovation and Information Flow*, Institute for Industrial Research and Standards.
15. —— (1982b), "Theoretical Implications of Measuring the Communications Sector", in M. Jussawalla and D.M. Lamberton (eds.), *Communication Economics and Development*, Pergamon.
16. —— (1982c), "Telecommunications in the Development Process", in M. Srinivasan (ed.), *Technology Assessment and Development*, Praeger.

17. —— (1983a), "Information Economics and Technological Change", in S. Macdonald, D.M. Lamberton and T.D. Mandeville (eds.), *The Trouble with Technology*, Frances Pinter and St. Martin's Press.

18. —— (1983b), "National Policy for Economic Information", in D.W. King, N.K. Roderer and H.A. Olsen (eds.) *Key Papers in the Economics of Information*, Knowledge Industry Publications.

19. —— (1983c), "Information, Organization and Development Policy", *The Information Society*, vol. 2.

20. —— (1984a), "Australia as an Information Society – Who Calls the Shots?," *Search*, vol. 15.

21. —— (1984b), "Exogenous Factors in Economic Theory", *Prometheus*, vol. 2.

22. —— (forthcoming a), "The Economics of Information and Organization", in M. Williams (ed.), *Annual Review of Information Science and Technology, vol.* 19, Knowledge Industry Publications.

23. —— (forthcoming b), "The Emergence of Information Economics", in M. Jussawalla and H. Ebenfield (eds.), *New Perspectives on Information and Communication*, North-Holland.

24. —— (forthcoming c), "Australian Regulatory Policy", in Marcellus Snow (ed.) *Economic of Telecommunications, Information, and Media Activities in Industrial Countries*, North-Holland.

25. LAMBERTON, D.M., MACDONALD, S. and MANDEVILLE, T.D. (1980), "The Word Processor Industry in Australia", in *Technological Change in Australia*, Report of the Myers Committee of Inquiry, Australian Government Publishing Service.

26. —— (1982), "Productivity and Technological Change", *Canberra Bulletin of Public Administration*, vol. 9.

27. LAMBERTON, D.M. and MANDEVILLE, T.D. (1983), *TBDF and Large Australian Companies*, Report for OECD, University of Queensland.

28. MACDONALD, S. (1975), "The Progress of the Early Threshing Machine", *Agricultural History Review*, vol. 23.

29. —— (1983a), "The Patent System and the Individual Inventor", *European Intellectual Property Review*, vol. 5.

30. —— (1983b), "Faith, Hope and Disparity – An Example of the Public Justification of Public Research", *Search*, vol. 13.

31. —— (1983c), "High Technology Policy and the Silicon Valley Model", *Prometheus*, vol. 1.

32. —— (1983d), "Agricultural Improvement and the Neglected Labourer", *Agricultural History Review*, vol. 31.

33. MACDONALD, S., LAMBERTON, D. and HODGE, B. (1981), "Tradition in Transition – Technological Change and Employment in Banking",

Working Paper No. 33, Department of Economics, University of Queensland.

34. MACDONALD, S., LAMBERTON, D.M. and MANDEVILLE, T.D. (1983), *The Trouble with Technology"*, Frances Pinter and St. Martin's Press.

35. MACDONALD, S. and MANDEVILLE, T.D. (1980), "Word Processors and Employment", *Journal of Industrial Relations*, vol. 22.

36. —— (1984), "Telecommunications in ASEAN and Australia", *ASEAN-Australia Economic Papers*, No 5.

37. MACDONALD, S., MANDEVILLE, T. and LAMBERTON, D. (1980), "Computers in Small Business in Australia", *Industry Economics Discussion Paper No.* 14, Department of Economics and Institute of Industrial Economics, University of Newcastle.

38. —— (1981), "Telecommunications in the Pacific Region – Impact of a New Regime", *Telecommunications Policy*, vol. 5.

39. —— (1982), "The Cost of Merit in University Research", *Australian Journal of Education*, vol. 26.

40. MANDEVILLE, T.D. (1983a), "Australian Use of Patent Information", *World Patent Information*, vol. 5.

41. —— (1983b), "The Spatial Effects of Information Technology", *Futures*, vol. 15.

42. —— (forthcoming) "Information Flows between Australia and Japan", *Papers of the Regional Science Association – Eighth Pacific Regional Science Conference.*

43. MANDEVILLE, T.D., LAMBERTON, D.M. and BISHOP, E.J. (1982), *Economic Effects of the Australian Patent System* (2 volumes), Australian Government Publishing Service.

44. MANDEVILLE, T.D. and MACDONALD, S. (1980), "Reflections on the Technological Change Debate in Australia", *Australian Quarterly*, vol. 52.

45. —— (1981), "Computers and Employment in Local Government", *Preprints of Papers of First National Local Government Engineering Conference*, Adelaide.

46. MANDEVILLE, T.D. MACDONALD, S. and LAMBERTON, D. (1980), "The Fortune-Teller's New Clothes – A Critical Appraisal of IMPACT's Technological Change Projections to 1990/91", *Search*, vol. 11.

47. MANDEVILLE, T.D., MACDONALD, S., THOMSON, B. and LAMBERTON, D. (1983), *"Technology, Employment and the Queensland Information Economy* Report to the Department of Employment and Labour Relations.

INDEX